EXOTIC FRUITS & VEGETABLES

EXOTIC FRUITS & VEGETABLES

Text by Jane Grigson
Illustrations by Charlotte Knox

JONATHAN CAPE
THIRTY-TWO BEDFORD SQUARE LONDON

To Joe
and in memory of Geoffrey

We would like to thank the staff at the Herbarium of the Royal Botanic Gardens, Kew, and Caroline Whitefoord of the Botany Department, British Museum (Natural History), for helping to identify various of the more obscure gourds and beans.

First published 1986

Jonathan Cape Ltd, 32 Bedford Square, London WC1B 3EL

British Library Cataloguing in Publication Data

Grigson, Jane
Exotic fruits and vegetables.
1. Cookery (Fruit) 2. Cookery (Vegetables)
I. Title II. Knox, Charlotte
641.6'4 TX811

ISBN 0-224-02138-9

Printed in Italy by New Interlitho SpA, Milan
Photoset by Rowland Phototypesetting Ltd,
Bury St Edmunds, Suffolk

CONTENTS

INTRODUCTION

In January 1981 Charlotte Knox walked into the offices of Jonathan Cape, in London, with a portfolio of drawings – coloured drawings of exotic fruits, vegetables and spices, depicted with a clear intense beauty and meticulous accuracy. Her idea was to make an album in nineteenth-century style, with plates vivid enough for people to be excited by them, to want to pick them off the page and try them for themselves.

The fruits and vegetables she has chosen, all from the world's hotter climates, are beginning to appear in markets and small grocery stores all over the country, wherever there are immigrant communities homesick for their native food. Some of the items are now quite well established in the better supermarkets – lemon grass, okra, persimmons, prickly pears, lychees, mangoes and white radish have become quite familiar sights in Wiltshire, though as a rule I go straight to a small Indian shop in Swindon, open to all hours, when I want something out of the ordinary.

The snag about buying unusual things in these stores is language. You cannot acquire any very exact knowledge of how to proceed apart from a general idea that most vegetables can go into a curry. We hope that this book will provide information and encouragement. A lack of information can be most off-putting. Even the bravest cook will find her or himself going home with cabbage instead of karela. We really have no idea of the glorious array of vegetables and fruit imported into this country, the vast undercover trade with ethnic restaurants, shops and markets, that makes few dints in the general laziness of greengrocery as we have always known it. It is time shopkeepers woke up and tried a little missionary work. They might find life more entertaining – their customers certainly would. Everything illustrated in this book has been bought in this country, admittedly in London though it could well have been found elsewhere, if only we made more demands on our retailers.

Two great Aladdin's caves of unsuspected glory are to be found just off the M4, not far from Heathrow. The Western International Market has a fruit and vegetable department of dazzling resource. Of course it is a wholesale market, but there is no reason why you shouldn't go and take a look. Not far away in Southall is the cash-and-carry owned by Mr Farouk Suterwallah, a great warehouse and emporium of dull aspect – until you begin to read the labels. Spices, special foods, strange kinds of salt, bags of poppadums, sacks of pepper, equipment for various kinds of oriental cookery, go from floor to unreachable heights. After a bemused visit, we went to have a chat with Mr Suterwallah. He spoke lovingly of rice – rice has its years, like wine, something I had not suspected. It is not just short or middling or abundant as a crop, it has a finer or coarser flavour, better or worse cooking qualities, and so on. I learned that there is much discrimination involved in the buying, that the supermarkets supplied by Mr Suterwallah get the very best basmati and other rice: all the same, I determined to buy rice from small traders selling it loose to see if I could indeed discriminate between rice from different years. At least one item in a cookery book I have was properly explained – '1½ lbs of old rice' – did it mean old and dusty, old and stale or old and hard? I now take it to mean the best rice, just as old claret doesn't mean the dregs or leftovers but the best *premiers crus*.

Another thing Mr Suterwallah told us about, and a beautiful picture he made of it, was the manufacture of poppadums. All round the factory where the dough is made and shaped is a great drying ground, spread with cloths. Girls in bright saris flit from cloth to cloth dealing out poppadums like cards, so that they can dry out properly in the sun. When the weather seems on the turn to the monsoon, they all flit down to the south of the country, the whole factory staff, and spread out their cloths again in the sun. A few turns of the wrist and the poppadums are drying with hardly a break in production.

There has been such an increase lately in good ethnic restaurants, and in helpful television programmes like Madhur Jaffrey's series, that most of us can begin to see how to use spices in individual combinations, according to the vegetables we want to try out. Down with the general yellow–brown curry powder! Fruit is another matter; occasionally it may take its place in a savoury dish, duck with lychees for instance, or pawpaw stuffed with a spicy meat and rice filling, but we have not quite reached the general European idea of fruit as the finale to a meal. In Britain we still cling to our puddings, which are often rather stuffy. This is bound to change; as concern with healthier eating takes effect, puddings and glamorous cakes will be relegated to Sundays and special feasts and we shall follow the European style of cheese after the main course, then fruit to finish.

I have the feeling that the giver of dinner parties has an idea that she – or he – should slave away at every course. Otherwise they are getting off too lightly. I read somewhere that the good hostess always offers a choice of two or even three desserts. Plain ridiculous. In New York, at the Café des Artistes, I found just the answer to dessert at the end of a special meal. We had been eating brunch, that strange Sunday meal of America, beginning with the typical Bloody Mary. I was intrigued to see how it would be rounded off. The waiter put a large and elegant dish on the table, an old meat plate I would say, and on it, splashed like brush strokes in a Matisse painting, were sections of pineapple complete with leaves and skin, the flesh sliced and cut in chunks but left in place, wedges of melon, long pieces of kiwi fruit, a few black grapes, chilled quarters of fig in a hollowed-out grapefruit skin, pieces of pawpaw sprinkled with lime juice. You could have added mangosteens, half-peeled rambutans and lychees. Each person was given a fork, and as we talked, feeling relaxed at the end of the meal and having come to know each other over the excellent food, we speared bits of this and that, recommended anything particularly good to our neighbours. With a dessert wine, this was something anyone could serve. All that was needed was a fine big dish and an eye for the glory of the fruit.

Precisely the same thing can be done with vegetables, the harvest festival on a single dish, mixing familiar with exotic items, binding the whole thing with a single sauce. The Indonesian salad gado-gado (p. 49) is a good example, with its peanut sauce, or the Provençal *aioli garni* of warm and cold vegetables, salt cod, snails and a vast bowl full of a very garlicky mayonnaise. Sometimes in the spring you can mix new young vegetables, including a few exotics, and serve them with an hollandaise sauce. Many of the recipes in this book are necessarily for foreign dishes, but in the end the assimilation of new vegetables depends on absorbing them into our normal ways of eating. The occasional stab at an authentic dish is fun. What really matters, though, is finding where new discoveries fit in, how they can enliven our diet without distorting its balance or making it seem bizarre. Odd to think that once upon a time the potato or the tomato was as strange to our ancestors as the bitter gourd or the roselle is to us.

The two best pieces of advice I have ever had about food came from unexpected sources, and they both apply to the contents of this book. My mother, when I was a child, said, 'Always choose from a menu something you have never tried before, and if you see a new fruit or vegetable, buy it to see what it tastes like!' Much later on, when I started writing about food, rather nervously, a close friend in France, who had mostly led a vagabond peasant life, was talking about the composition of the best Italian salami (which includes donkey meat, or did at that time before EEC regulations). I said, 'Ugh!' He turned to me gravely, with a reproof. 'If people from a different culture value a food, you may be sure they do so for a good reason and you should try it without disgust.' It is no hardship, I assure you, to try the delights of this book.

Quantities and Measurements

Generally speaking the recipes in this book are for six people. However, many of the dishes come from cuisines in which the food is deployed in buffet style, as it was in Europe until Escoffier came down firmly in favour of the Russian service – one dish at a time and a number of courses *seriatim*. Eighteenth-century cookery books often have diagrams of the two or three courses served at a dinner party. They are set out rather like a Chinese meal, though of course the actual food is different. This style has the advantage of elasticity. If an unexpected guest arrives, you do not have a nervous crisis because there are only six chops and he makes seven at table – you quickly add another small dish of some store cupboard item, or two dishes if you are feeling lavish.

Often, too, quantity depends on income. In rice-eating civilizations the other items are seen almost as devices to tempt the staple food down. If you are poor, you have less temptation. The closest equivalent in our West European eating is the French style: I notice in our district that everyone buys good things from the same good shops, but the road-mender's wife buys less and serves more bread with it than the Ambassador's wife. This seems to me a better solution (as long as you can rely on the bread), than the less well-off buying large quantities of cheaper, and often trashier, food. Now that we have such a wide choice of different kinds of rice and beans, and a much better choice of vegetables thanks to our immigrant communities, we can stretch our money further and better by eating more of the starchy staples.

Measurements are in metric and imperial. Follow one system or the other, don't jump in the middle of the recipe. All spoon measurements are level, unless otherwise specified. I use a 15 ml (½ oz) tablespoon, the equivalent to three 5 ml teaspoons: if you do not have a set of measuring spoons, make a resolution to buy one and meanwhile use the plastic teaspoon given you with bottles of medicine, which is the 5 ml size.

For purposes of space, I have had to assume that every reader has a basic knowledge of cookery, or at least a basic cookery book to refer to. Remember that recipes are always in the end suggestions, and not laws to be obeyed: you are dealing with things that have grown, their development affected by climate and man, new varieties are always coming into cultivation, familiar fruits are being imported from unfamiliar sources. That is the fun of cooking and eating new things.

Finally, the fruits and vegetables in this book all have a variety of regional names. The more common ones are listed in the Glossary on pp. 122–4.

FRUITS

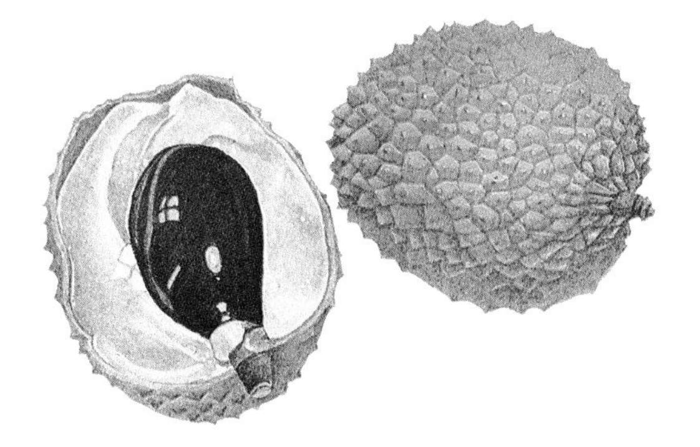

CARAMBOLA

(Averrhoa carambola)

Carambola is a waxy yellow fruit that is easy to recognize from its five ridges. An Italian monk travelling in the East Indies wrote in 1672 that it is 'as large as a pear, all sculptured (as it were) and divided into ribs, the ridges of which are not round but sharp, resembling the heads of those iron maces that were anciently in use'. If you slice it across, it falls into a cluster of decorative yellow stars – hence its other names of star fruit or star apple – and has a taste that falls sweet and sharp on the tongue. This tropical acidity groups it with those other useful items such as lemon, tamarind, rhubarb, sorrel, gooseberry, that with a little guile and ingenuity on the part of the cook can be interchanged as seasonings. Not surprising to learn that it is related to wood sorrel, whose juicy sharp leaves are much relished by children in the spring. Carambola, originally the Portuguese name, goes back to the Sanskrit *karmara* meaning 'food-appetizer'.

To the English living in southern Asia the carambola was known as the Coromandel gooseberry. And to settlers in southern China it was obviously the Chinese gooseberry. They were told that the Chinese name for it was *yang t'ao*, meaning goat peach. And when they heard Chinese from the north using the same name for the furry *Actinidia* (kiwi fruit, see p. 18) they simply transferred their English name for the carambola to the more northerly-growing kiwi. As Professor Edward Schafer says, 'The term Chinese gooseberry should be expunged. Meanwhile remember that a request for it may well get you a kiwi fruit in London, but a carambola from an English-speaking merchant in Hong Kong.'

When the carambola is green and the astringency is most marked, it is used as a vegetable and for pickles and chutneys. Since it is native to Malaya, and spread long ago to southern China and India, it is from the cookery books of these areas that you will get the best ideas, such as Padang Sour-Sharp Fish on p. 13. For this you really need the tiny sharper bilimbi (*Averrhoa bilimbi*, which have a similar shape), although you can use unripe carambola instead: be guided by flavour.

To Prepare

Choose carambola that have not begun to show brown patches. Should the sharp angles of the fruit be marked with a brown line, pare it away delicately. Then cut the fruit across, usually in slices about ½cm (under ¼in) thick.

Padang Sour-Sharp Fish (Pangek Ikan)

The subtlety of this recipe is in the three acidities of lemon, lemon grass and star fruit. Macadamia nuts are to be found in good grocers and some health-food shops: as a last resort, try Brazil nuts, which have a similar waxiness.

7 macadamia nuts or 4 Brazil nuts
2 medium onions, sliced
2 large cloves fresh young garlic
1 level tbsp fresh ground chilli, including seeds
2 ½-cm (less than ¼-in) slices ginger root, peeled
¼ level tsp turmeric
½kg (1 lb) sliced fresh tunny, bonito, grouper or monkfish
1 tsp salt
2 tbsps lemon juice
1 stalk lemon grass, bruised
6 small bilimbi or 3 unripe carambola or ¼ lemon cut in thin wedges

Chop or process the first 6 ingredients to a smooth paste. Set the fish to marinade for ¾ hour in the salt and lemon juice. In a heavy pan that will take the fish in a single layer, put the paste, 4 tbsps water and the lemon grass. Bring to simmering point, slice and slip in the bilimbi or carambola, or add the lemon wedges. Cook gently for 5 minutes, stirring often to prevent sticking. Put in the fish, turning it over so that it is coated, then cook it until just tender (about 10 minutes). Shake the pan gently and turn the fish carefully so that the slices do not collapse. Remove the pan from the heat and cool down. Cover and leave until next day. Reheat carefully (which is why you should avoid overcooking the fish in the first place – it continues to cook as it cools down).

Mango and Carambola Salad

Carambola makes the most delicious syrup for fruit salads, and you have the bonus of the starry shapes as well.

300g (10oz) sugar
2–3 carambolas, sliced
3–4 fine ripe mangoes, peeled, stoned, sliced
gin (optional)

Make a light syrup by dissolving the sugar in 600ml (1pt) water over a low heat, with the ends of the carambola. Bring to the boil and boil steadily for 4 minutes. Slip in the star slices of carambola, lower the heat and simmer until they are tender (about 7 minutes). Remove the best stars for decoration. Taste the syrup and regard its consistency. If it needs to be more strongly flavoured and thicker, boil it down hard again. When you judge that it is about right, put in the mango and bring back to the boil. If the fruit is on the firm side, give it a minute or two's cooking: if it is juicy and just right, remove the pan from the heat when it returns to the boil. The idea is to impregnate the mango with some of the acidity, not to cook it in the accepted sense.

Arrange the mango slices in a shallow dish with the stars of carambola you set aside. Taste the syrup as it cools down and decide whether a little gin would give it an agreeable lift. If so, add it by the tablespoonful. Gin is a splendid spirit with fruit, sometimes to be preferred to kirsch.

Strain over the mango, being careful not to drown it. Remaining syrup can be stored in the refrigerator for other fruit salads.

Note: ripe orange-fleshed melon can be substituted for mango. With pears, be prepared to cook them a little longer. In a mixed fruit salad – be discreet with the mixture – some of the fruits will not need cooking or heating at all, for example lychees.

CHERIMOYA, SOUR-SOP and SUGAR APPLE

(Annona)

There are about sixty species of *Annona*, but we shall be lucky – most of us – if we come across a handful of them. The cherimoya (*Annona cherimolia*) is smooth and green, the size of a pear usually when we see it in the market, with a pattern that looks like the green scales of a young pine cone at first glance. The sour-sop is prickly outside with soft spines in green rows, and is a much larger affair: the soft flesh of the *Annona* takes on an acidity in the sour-sop that makes it an excellent base for drinks (seed and remove the pulp from the shell and liquidize it with milk or water to taste), or a fruit fool with whipped cream, or sherbet or sorbet. But this creamy facility is a characteristic of all the *Annona* fruits we eat, each one tasting a little different, whether it be the sweet-sop or sugar apple (see p. 124), which is particularly sugared in its flavour, or a hybrid like the atemoya, whose parents are the cherimoya and sweet-sop.

I suppose we shall see more hybrids and cultivated varieties in the future, labelled with unpleasant commercial names – African Pride, for instance – which tell you nothing. For my part I prefer the old names for fruit which Mark Twain described as 'deliciousness itself'. Cherimoya, for a start, which comes from the Peruvian *chirimuya*, meaning cold seeds. Indeed its smooth black seeds have been found in Inca graves, and so have Chimu and Nazca pots shaped like the fruit itself. Cold seeds is just right.

If you can afford on special occasions to slice the top off a cherimoya like an egg, and pour in a little cream, that is the best way to eat it. Just watch out for the seeds, those cold black seeds. And be careful when you bring the fruit home from the shop: their skin is tender, it can be easily broken and the pulp crushed. On not so rich occasions, you may want to stretch the pulp by turning it into a sorbet, ice cream or soufflé.

Cherimoya
(*Annona cherimolia*)

Cherimoya Soufflé

200–250 g (6½–8 oz) pulp, weighed without seeds or skin
lemon juice
4 egg yolks
175 g (6 oz) caster sugar
1 pkt (½ oz) gelatine
300 ml (½ pt) whipping cream
3 egg whites

Taste the pulp and add just enough lemon juice to emphasize the flavour. Whisk yolks and sugar in a bowl over a pan of water that has just boiled, until the whisk forms ribbons when it is taken up from the mixture and trailed across the surface. Dissolve the gelatine in 6 tbsps of very hot water and stir into the egg mixture. Whisk the cream until it is just firm but not stiff and fold it into the cooled egg yolk, with the pulp. Taste and adjust seasoning if necessary. Go lightly. Whisk the whites and fold them in. Turn into a mould and serve, if you like, with a sauce of sieved, sweetened strawberries.

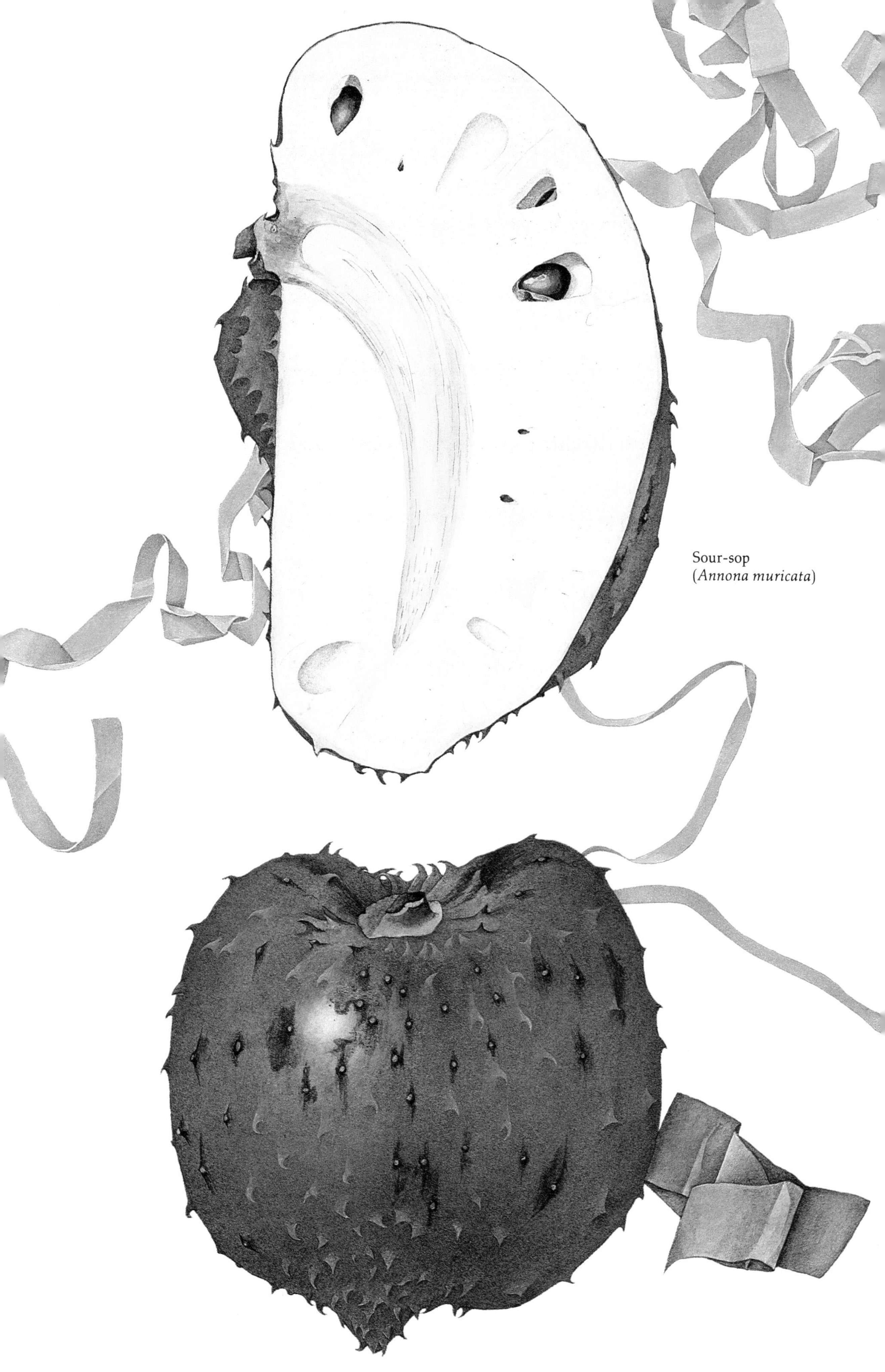

Sour-sop
(*Annona muricata*)

GUAVA

(Psidium guajava)

The guava, it must be admitted, smells better than it tastes (or looks). Like the quince it will perfume a room with its air of indescribable sweetness that to a northerner means some paradise in the tropics. And northerners, since they first described it in the sixteenth century, have tried to find parallels for its fragrance in our colder experience. Quince of course, not because the smell is in the least similar but because of its dominating power, which overrides but yet includes the sweetness of any other fruit that happens to be near. Fig and peach are other comparisons for this exotic *guayaba* – the name Columbus and his people learnt from the Arawak Indians of the West Indies. Hence guava and the French *goyave*.

The first account of the fruit, in 1526, describes its many seeds, or 'more properly speaking, it is full of hard small stones, and to those who are not used to eating the fruit these stones are sometimes troublesome'. By these stones or seeds the guava has been spread all over the hot parts of the world by birds. And outside the tropics, too. Once in Crete we were given some guava jam to eat, most delicately flavoured with ginger. The friend who had made it told us that a South African had brought a guava tree to the island, and the birds did the rest. Guavas are not sold in the markets there, but they grow in a few gardens. Nothing to compare with their appearance, say, in India, where one traveller in Oudh remarked that they found large orchards of wild guava 'strongly resembling in their rough appearance the pear-trees in the hedges of Worcestershire'.

Like a number of other fruit, guavas come in varying shapes – pear-shape, or round, or a blunted oval – and colours from green to yellow. When they are cut across, the compartments of flesh containing the seeds become apparent. Their stoniness makes a strange contrast with the musky, voluptuous flesh, which, in my experience at least, requires some acidity either from passion fruit, lime or lemon juice. People who have been lucky enough to eat them from the tree may not be of the same opinion. My own feeling is that guavas, apart from jelly and goyabada, are best used for flavouring syrups for fruit salads and ices. Cut them up, skin and all, and simmer them in a covered pan with sugar and water to extract the exquisite flavour: then strain out the fruit and soak or poach the fruit for the salad in the syrup, as appropriate.

Guava Jelly

No need to peel. Slice about 1 kg (2 lbs) fruit into a pan and add water to cover. Simmer until the guavas are very tender, then strain off the juice, pushing through some of the pulp if you are not obsessed by jelly with a jewel-like clarity. Alternatively keep the pulp to make goyabada (guava cheese – see p. 17). Take 600 ml (1 pt) of the liquid, sweeten it to taste, bring out the flavour with a little citrus or passion-fruit juice, and infuse over low heat with a few crushed cardamom seeds or some bruised fresh

Guava
(Psidium guajava)

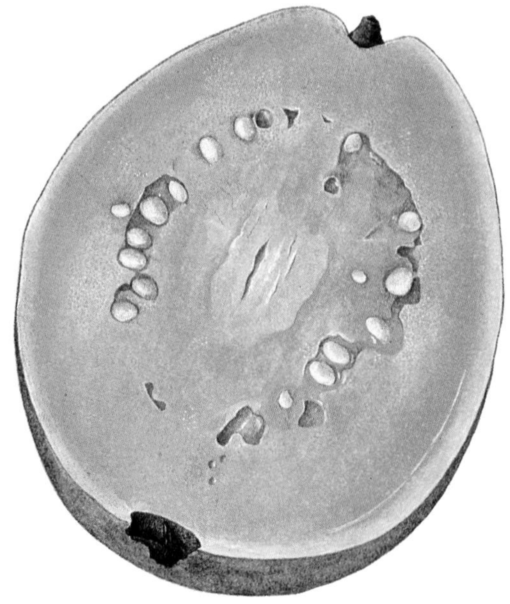

Pear-shaped guava
(*Psidium guajava pyriferum*)

ginger to give a light spicing. Do not overdo this. Dissolve 1 pkt gelatine in 5 tbsps of very hot water and stir in the guava liquid. Pour into 6–8 glasses and chill to set. Serve with a layer of cream on top, or with coconut cream (see p. 39). A few chopped macadamia nuts on top make a good contrast, or some shavings of Brazil nut.

Guava Purée

Peel and chop ½ kg (1 lb) guavas. Put them in a pan with enough water to prevent sticking. Add grated zest and juice of a lemon and 8 level tbsps sugar plus a few cardamom pods or ground or grated fresh ginger, if you like the idea. Cook just until the guava is soft enough to be pushed easily through a sieve. Taste and add extra sugar or lemon juice as required, and according to the destination of the purée. Use to make soufflés, ices or a fool. If you keep the sugar content low, you can serve it as a sauce with pork or duck.

Guava Liqueur

Peel and seed enough guavas to fill a 1 kg (2 lb) bottling jar. Put peel and seeds into a pan with a breakfast cup of water and 2 of sugar. Stir over a low heat until dissolved, bring slowly to boiling point and boil for 2 minutes. Strain and cool. Pour enough syrup into the jar to come just over a third of the way up the fruit: keep the rest of the syrup in the fridge. Top up the jar with brandy. Cover and leave to mature. Taste after 2 or 3 months, longer if you manage to forget it and avoid temptation, and add more syrup or more brandy after straining off the liqueur.

Goyabada

A Brazilian favourite, though this is made all over Latin America. The method is the same as we all use in Europe to make fruit pastes such as Spanish *membrilo* and French *cotignac* from quinces. Cook the fruit, having quartered it, with water just to cover. Then sieve it, or process and sieve it, and weigh the pulp. To each 500 g (1 lb) allow 500 g (1 lb) sugar. Heat gently together in a heavy pan until the sugar dissolves, then raise the heat and stir the mixture until it is so stiff that a wooden spoon can make trails on the bottom that are not instantly filled. If you are not used to making fruit pastes, take particular care to wrap your hands in cloths towards the end because the pulp bubbles up and bursts with an explosion of boiling fruit that can burn you. Line a shallow oblong tin with butter muslin after brushing it thinly with tasteless oil (sunflower or groundnut, for example). Pour in the purée to a depth of 1 cm (½ in) and lay butter muslin over the top, pressing it flat. Dry the paste off in an airing cupboard or over the stove for up to 2 days, until you can cut it into slabs or small squares. Roll and store in granulated sugar if you like, or wrap in cling film, then foil, and keep in a cool place. Goyabada can be served as a sweetmeat with coffee, or as a dessert with soft cheese, or with grilled lamb cutlets or other meats.

KIWI

(Actinidia chinensis)

Of all the fruits which have come to us from China, I suppose the kiwi or Chinese gooseberry is the most recent. It is also one of the best, not just for its cool, elegant appearance when sliced and its refreshing flavour, but also for its vitamin C content, which is higher even than a large orange's. And it is low in calories. We have every reason to eat it with enthusiasm. The odd thing is that it has not been eaten with enthusiasm by the Chinese – this, I presume, is why it has arrived so recently in the West by comparison with the orange and the peach, which have brightened our tables for hundreds of years.

As Chinese gooseberries the fruit reached Europe nearly a century ago and seeds of the plant were tried out in New Zealand, where the first crop was harvested in 1910. Progress was slow. I remember eating a kiwi fruit once as a child before the Second World War, but how it had reached our house in the north of England I do not know. The first shipment abroad – to London – was not until 1953. And it came with the new name of kiwi. It is not a name I like. It reminds me of shoe polish. The old name recalled the origin of the plant, its translucent green flesh and the puzzlement of Europeans who first encountered it in the East. But unfortunately they had already given the name of Chinese gooseberry to the carambola or star fruit (p. 12). So reluctantly we have to accept kiwi as its name. After all New Zealand has done all the work.

Nowadays fruit farmers in California and France, and other countries, are setting up their T-shaped pergolas by the mile and the kiwi becomes more popular every season. At first it was the chic fruit of top chefs. It is now chic to sneer at it in a knowing way, because second-rate chefs drag it into every dish and have made it a cliché for inappropriate garnishing. Luckily the general public have the sense to buy and enjoy kiwi fruit because they are easy, delicious and virtuous eating, with a good storage life unusual in a soft fruit.

To Prepare

Kiwi fruit can be cut in half either way and eaten with a spoon. Or they can be thinly peeled with a potato peeler or sharp stainless knife, and then cut into slices, or down into quarters or wedges. Judge ripeness by cradling them in your hand and giving them a gentle squeeze – if they give a little they are ripe, if they are firm put them into a paper bag with an apple or banana or keep them in a warm room for a day or two. If you want to cook them, think of them as grapes or very fine gooseberries: heat increases their acidity.

Pavlova

Kiwi and passion fruit are the standard flavourings of a Pavlova cake, which was

first made in honour of the dancer when she visited Australia in the 1930s. The soft meringue is, I would say, by origin European, but the shape it is baked in and the filling are Australian.

3 large egg whites
175 g (6 oz) vanilla-flavoured caster sugar
1 level tsp cornflour
1 tsp wine vinegar
250 ml (8 fl oz) whipping cream
2 large passion fruit
extra sugar
4 kiwi fruit, peeled and sliced

Using an electric beater if possible, whisk the whites stiff. Add the sugar, beating until the meringue looks silky. Fold in the cornflour and vinegar. Pile or pipe on to a baking sheet lined with vegetable cooking parchment, in a nest shape about 25 cm (9 in) diameter. Bake 1¼–1½ hours at gas mark ½, 130°C (250°F), or until firm. When cool, peel off the paper and put the nest on to a plate. Whisk the cream, sugaring it to taste. Just before serving, pile the cream into the nest, spoon the passion fruit over it, and arrange the slices of kiwi fruit on top.

Kiwi Gratin or Tart

Peel and slice 8 or 9 kiwi fruit. Either butter a gratin dish, or bake blind a shortcrust pastry case until set but not coloured. Beat 4 egg yolks with 4 tbsps water and 50 g (scant 2 oz) vanilla-flavoured caster sugar, using an electric beater, until very bulky and light. Whisk 250 ml (8 fl oz) double cream until stiff and fold into the egg yolks. Put the kiwi slices into the dish or pastry, sprinkle them with 4 tsps kirsch or gin, and pour over the egg mixture.

Either poach the gratin in a *bain marie* for 20 minutes, then glaze under the grill. Or bake the tart in the oven preheated to gas mark 4, 180°C (350°F) for 20–30 minutes. Sift the top with icing sugar, protect the pastry with a circle of paper and put under a red-hot grill for a moment or two to glaze.

Kiwi and Tamarillo Salad

Tamarillo, being slightly tart, makes a good partnership with kiwi. Allow equal quantities of each, according to the number you intend to feed. Peel and cut the kiwi fruit into slender wedges. Scoop out the flesh of halved ripe tamarillos and chop it roughly. Arrange the two together in a shallow dish. Sprinkle with caster or icing sugar – a level tsp for each pair of fruit is about right.

Kiwi Fruit and Chocolate

Kiwi fruit is refreshing with chocolate, a sliced kiwi salad with a chocolate mousse for instance, or kiwi slices and whipped cream to fill a chocolate cake.

LYCHEE

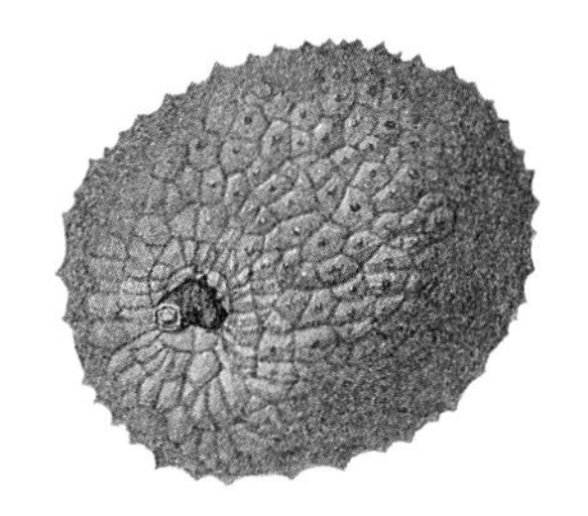

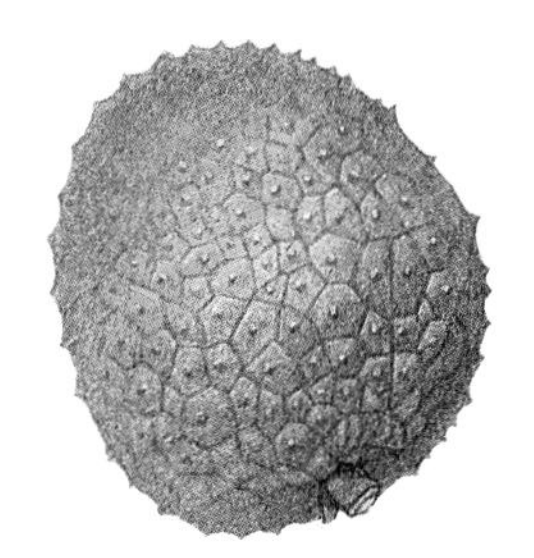

(Litchi chinensis)

Of all tropical fruit, the lychee has acquired the most sensuous and emotional wrappings in its long history of giving pleasure. I would say that this is not only because of the scented flavour of the fruit, but also because of its shape and white translucency. In China where it comes from it was the supreme fruit, so prized that it was taken to the northern court from hot southern Kwangtung, where it grew, by swift horses. This extraordinary courier service, running as early as the Han dynasty and involving much loss of life, went on for centuries. Above all it is associated with the Lady Yang Kuei Fei, the Precious Consort of the great T'ang Emperor Hsüan Tsung, who ruled from AD 712–56:

Looking back at Chang'an, an embroidered pile appears;
A thousand gates among mountain peaks open each in turn.
A single horseman in the red dust – and the young Consort laughs
But no one knows if it is the lychees which come.

They were her favourite fruit, just as peonies were her favourite flowers. 'A peony looking at a peony' is how she was once described – a vivid image of this slightly plump, voluptuous beauty, whose skin was flushed with rose under its pale colour, gazing at perhaps a rose-scented peony in full pink and creamy bloom. She was talented as a poet and graceful as a dancer. And she had, as it were, her Shakespeare in Po Chu-i, one of China's grandest poets, who wrote her story in his poem 'The Everlasting Wrong'.

The trouble was that Lady Yang did not confine herself to love and chewing lychees. For over twenty years she ruled the Emperor's judgements as well as his emotions. Her influence was resented, her family's advance envied. In the end the royal lovers were cornered by the Emperor's own bodyguard. Lady Yang was forced to hang herself from an old pear tree. The Emperor watched as her jade hair pins and gold ornaments fell to the ground.

When after two years exile Hsüan Tsung was allowed to return, he was brought a scent-bag dug up from her roadside grave. It smelled sweetly of her still, and he wept. No doubt he remembered the summer days when they had cooled their mouths with melon served in iced jade bowls, and with lychees brought so swiftly from Kwangtung that their evanescent flavour was unimpaired. (The first fruit to be brought by refrigerated transport? But how did they have ice down in Kwangtung?)

A magician was dispatched to find the spirit of Lady Yang. She gave him old keepsakes, the Emperor's gifts, and told him the secret oath that they had sworn to each other on the seventh day of the seventh month, an oath that they would never be separated but fly in heaven, two birds with the wings of one, and grow on earth entwined like the branches of a tree. Heaven and earth shall pass away, said the Lady Yang, but the great wrong done to us will live for ever.

Now the seventh day of the seventh month is the festival of the Herd Boy and the Weaver Maid, very much the lovers' festival, in the middle of the summer, when in the south at least the lychee fruits hang pink and ripe on the elegant trees that often border canals and waterways. Fruit-sellers used to go round the streets crying, 'Juicy-juicy-sweet, full-full fragrant, sweet-smelling, red and watery fresh-peeled, round-eyed lychees from Fu-chou . . . Oh you sweet elegant ladies, beautiful women from fragrant chambers and embroidered kiosks, great and noble gentlemen from high halls and great buildings . . . try mine!' And a very similar fruit, smaller, less fragrant, the longan, was eaten by all the young girls. This was because longan means

dragon's eyes, and if they ate a lot of them perhaps their eyesight would improve enough to enable them to weave and embroider as beautifully as the Weaver Maid had done.

She was the daughter of the Sun, and her parents worried because she would do nothing but work at her embroidery and weaving. They found her a husband, a handsome Herd Boy. She left the web, she left the loom so completely, became so lazy in love, that her parents separated her from her husband, as you can see in the summer sky at night where the Milky Way separates Altair (the boy) and Vega (the girl) with a flow of stars. But her father had pity on them and once every seven years magpies form a bridge over the Milky Way for the lovers to meet. According to Doris Yen Hung Feng, in *The Joy of Chinese Cooking*, a favourite dish for the festival is a salad of seven fruits – watermelon, melon of another kind, oranges, cherries, peaches, grapes and dragon's eyes or lychees – served in a basket cut from the watermelon shell, with sweet wine and honey.

People have devised other ways of using lychees. The Chinese dry them to a raisin-like consistency – sold in oriental stores as lychee nuts – when they become rather dull. Chinese cooks, not I suspect of the top class, substitute them for pineapple or use both fruit together in sour-sweet sauces. Western cooks may open a can of lychees – not to be despised, since something of their texture remains – and whizz them into creams and sorbets. I have tried all these things more than once, and have concluded that they are all a waste of a lovely and sensuous fruit, almost an insult.

By far the best thing is to serve them on their own, still on their thin twiggy branches if you can, at the end of a friendly dinner. Then they can be admired as they should be over the pleasant chore of cracking the shells and peeling them away to reveal the plump, pearl-white fruit. In some varieties the shells are a deep pink, an extra beauty that caught the eye of a French poet on the Ile de Réunion as he watched the girl he loved being swung down to church in a hammock-litter known as a manchy:

Et tandis que ton pied, sorti de la babouche
Pendait, rose, au bord du manchy
A l'ombre des bois noirs touffus, et du
 letchi
Aux fruits moins pourpres que ta bouche.

A good image of pinkness and love, a pink foot that had slipped from its embroidered Turkish mule, a deep purple-pink mouth, deeper toned than the fruit on the lychee trees, all seen in the hot black shade of the leafy wood. A very different sensuousness from the Lady Yang's air of chilled fruit and jade, but the same fruit.

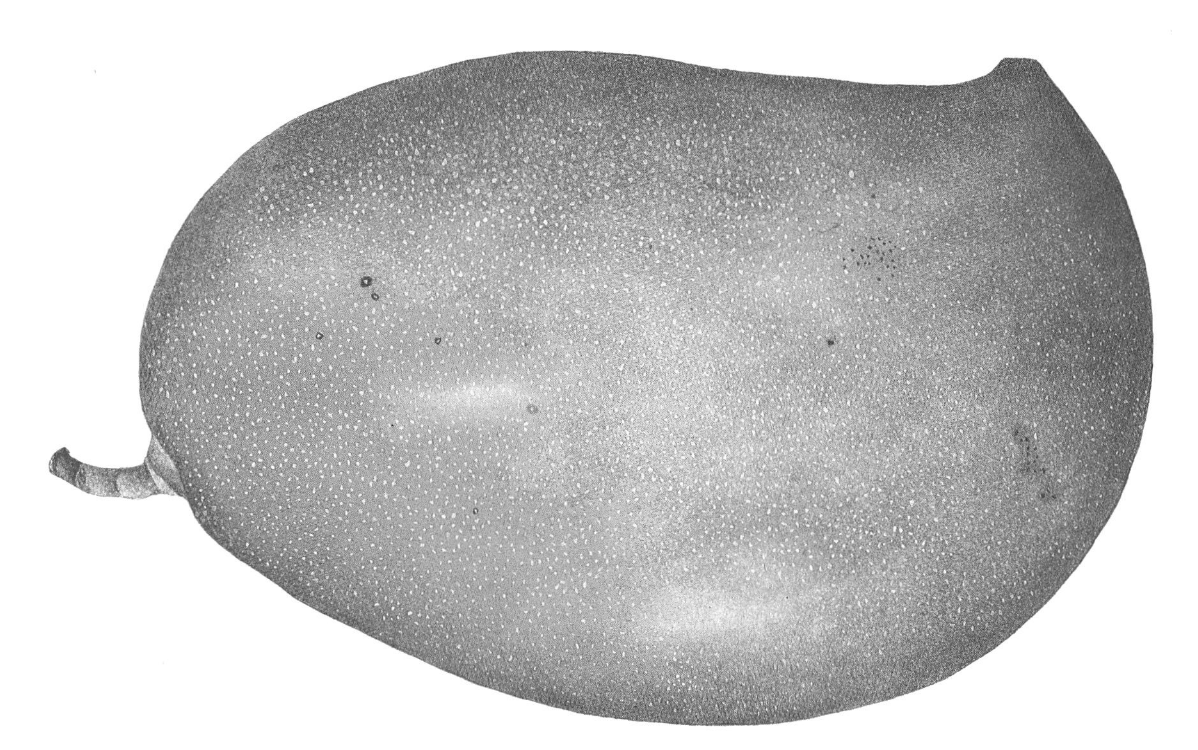

MANGO

(Mangifera indica)

Of all exotic fruit, perhaps of all fruit, the mango is the finest. Not every mango all the time, but the mango at its best. 'The mango is the pride of the garden,' wrote one great poet of the thirteenth century, 'the choicest fruit of Hindustan.' In the sixteenth century Akbar, the Mogul Emperor, who was a contemporary of Elizabeth I, loved mangoes so much that he had a great orchard planted entirely with them, the Lakh Bagh near Darbhanga, a hundred thousand trees.

Mangoes appear to have been cultivated for over 4,000 years, and in their natural state are little better than 'a ball of tow soaked in turpentine'. Reducing the fibre and bringing the turpentine down to an enticing breath that adds character to the richness of the fruit was the work of generation after generation of gardeners. By the time of the Buddha it was reckoned a fine enough fruit for a mango grove to have been given to him for meditation. The Portuguese took to it enthusiastically, judging by some of the names associated with the fruit, which they called *mangas* from the Tamil (it seems that mango came to us via the Dutch, from the same source). The finest type of mango for high dependable quality is the Alphonse, from Affonso d'Albuquerque, an early Governor of Portuguese territories in western India. The finest group of mangoes is the Mulgoba, meaning 'makes the mouth water', while the Sandersha ('parrot beak' from the sharper oval shape) is best for cooking.

But how are we to judge their quality when we regard a tray of mangoes in the fruit department of a supermarket? I do not understand why they have escaped the labelling regulations of variety and origin. Only when we know these things will we be able to discriminate among the many varieties and learn, as it were, to know the Cox's Orange Pippin from the Bramley and Golden Delicious. If only the early years of the introduction of this fruit could be better handled – and why not a Mango Council to look after this? – the mango could make a permanent part of our most delightful eating, could become the pride of our table. No longer just an embellishment as Dryden described it in talking of an unripe poet:

There's sweet and sour: and one side good
at least
Mango's and Limes, whose nourishment is
little,
Tho' not for food, are yet preserv'd for
Pickle.

To Choose

Choosing good mangoes is not easy. However carefully you consider them, you will find that they vary as much and more than any other fruit. The perfect mango, like the perfect peach, is an experience that may occur rarely and quite unexpectedly. Thanks to imports we can now buy them for most of the year – something that people living in India and Pakistan will envy us, since the mango season is a short one there. Families with mango trees in their gardens rarely get the chance of ripe fruit, since children start their secret thefts when the mangoes are green and acid. I confess I like mangoes with this fresh sharpness, when they are just right for jams and chutneys, for savoury dishes and for cooking lightly with some sugar. At this stage, as a rule, you will only find them in Indian stores and markets.

The mangoes found in supermarkets and general greengrocers are intended to be eaten ripe. If they are a little hard, keep them in a warm place for a few days until they begin to give when you cradle them in your hand and squeeze gently. Colour is nothing to go by. There are many varieties with differing appearances. Some are deep green with a red flush in places, some are plain green, some are yellow and some are that glorious sunburst of yellow and pinkish orange which seems to announce the perfect fruit.

To Prepare

At first preparing a mango can seem tricky. Stand near the sink so that you can rinse your hands and arms easily. First you should understand that the stone, though wide, is narrow in depth. This means that if you stand the mango upright, narrow side towards you, you can slice down a little distance from the stalk on either side – this gives you two shallow boats of mango, plus the stone with a ring of pulp and skin round it.

Once you get used to the mango's anatomy, you can slice in directly with a small sharp knife held at the right kind of angle to free the flesh from the stone, without making too messy a job of it. This gives you larger 'boats' with a concave top in which you can pour cream or coconut cream, or arrange pieces of lime. If the intention is to eat the mangoes out of hand, score the pulp of the 'boats' in a diamond pattern just down to the skin; the eater then picks up the piece and bends the skin back, pushing the centre part up so that the diamond opens like a flower, which makes it easier to eat – in theory at any rate. Mangoes can also be peeled, like a peach: score the skin down, spear the mango with a fork and strip away the sections. If the mango is soft and juicy, I would not advise this course of action. It would be better to slice in to the stone so that the pieces fall out in wedge-shaped sections, then remove the skin, or remove and eat the flesh from the skin with a knife and fork.

All in all, there is something splendid about eating a mango which is not for the prim and the pernickety; as the greengrocer realized when he put up a poster saying, 'Share a mango in the bath with your loved one.'

Mango Salad

Arrange slices of ripe, peeled mango on a dish. For every 2 mangoes allow 1 passion fruit. Put the passion-fruit pulp and seed in

Cooking mango

a small pan with water to cover. Bring slowly to boiling point then remove and sieve. Add caster sugar to the warm liquid to the sweetness you like and enough lime juice to bring out the flavour, then spoon this sauce over the mangoes. Scatter some of the black passion-fruit seeds over the top, with thin curls of green lime zest. Serve lightly chilled.

Mangoes and Coconut Cream

Mangoes taste delicious when served with coconut rather than straight dairy cream, as do some other exotic fruits. It is simplest to buy a tin of coconut milk; however, this can be very expensive, and you may prefer to use it to flavour cream rather than on its own. The solid coconut cream sold by many grocers these days is not quite so rich, but by dissolving it in hot water you get a very satisfying result. The less water you use the thicker the cream will be. To make coconut cream from the coconut itself, see p. 39: when you have finished with the first purée, return the coconut to the blender and pour on nearly boiling water. Whirl it up, then strain again. This process can be repeated. You then can mix the various liquids together to suit your taste. If you leave the mixture to stand in a glass jug, the richest part, the 'cream', will rise to the top, leaving the coconut milk below. Desiccated coconut can be soaked with hot water and squeezed out in a cloth, with repetitions to make weaker 'milks'.

N.B. The liquid you hear rattling about in a coconut is not coconut milk but coconut water: the milk and cream are made from the white coconut itself.

Mango and Pawpaw Tart

Pawpaw or papaya (see p. 28) can be used to extend half-ripe mango in chilled soufflés, pies and tarts like this one. Often a sharpening of lemon or lime is a good idea. In separate pans simmer the peeled, sliced flesh of 2 pawpaws and 2–3 half-ripe mangoes with a little water. Sieve the fruit and measure it. You need roughly 250 ml (8 fl oz) of each. Mix and sweeten to taste, sharpening with a little lemon or lime juice if necessary (depending on the acidity of the mangoes). Put into a pan with a large beaten egg yolk, 1 level tsp cinnamon and a little grating of nutmeg. Cook gently until the mixture thickens, like a custard: do not overheat or the yolk will curdle. Pour into a baked pastry case. Cover with a meringue of 2 egg whites beaten stiff and sweetened with 2 level tbsps caster sugar. Bake in the oven preheated to gas mark 3, 160°C (325°F), until nicely browned, 20 minutes or longer.

Am Ki Chatni

A recipe for fresh chutney which I was given by an American friend. Make it no more than 8 hours before the meal, or it will lose its freshness.

2 mangoes, firm and underripe, weighing about 500 g (1 lb) each
1 tbsp peeled, finely chopped fresh coconut
2 tbsps finely chopped fresh coriander
1 tbsp finely chopped fresh ginger root
1 tbsp salt
⅛ tsp cayenne or other hot red pepper

Peel and cut the mango flesh into 1 cm (½ in) cubes, discarding skin and stone. Put into a bowl and add remaining ingredients, stirring them gently with a spoon until well mixed. If not serving immediately (which is best), cover with plastic cling film and keep in the refrigerator.

MANGOSTEEN

(Garcinia mangostana)

I went into Fauchon's in the Place Madeleine in Paris one day via the fruit department. I always go in by that door because the moment you push it open paradise wafts towards you. No need for blue sea, palm trees, since the exquisite blend of guava and passion-fruit smells, pineapple, oranges, mango, cantaloup and charentais melons, figs, apples and perfect pears with tiny wood strawberries, provide an incomparable sweetness of distant places. The next move is to check on all the trays and punnets to see if there is anything I've never encountered before. There always is. On one visit it was a heap of purple-brown fruit, rounder than apples, but about the same size. They might have been carved from some dark unknown wood, each one finished with a top ruff of firm, rounded petal shapes. They had no scent at all. They played no part in the wafts of scent around them. I took advice – 'Mangoustans, Madame!' – and bought three for the family supper.

Away from the splendours of Fauchon, they looked drab. Sniffs all round. I said firmly, 'They are the most delicious fruit in the world!' The family thought I had been robbed.

With a sharp knife I cut into the firm shape, which was not as hard as it looked, and moved it round until a cap came off with those winged petal shapes. Inside was a jewel of plump, white, sectioned translucency, set in a case of glorious pink. As we admired it, the pink faded into a deeper deader tone, and I lifted out the small sections. Even thousands of miles from Malaysia, they tasted fragrant, unlike anything else, except that the flesh had the consistency of lychees.

The most delicious fruit in the world? Yes, perhaps, in Malaysia – at any rate, off the tree, or from a market-seller's plaited tray in Thailand. To someone living in the British Isles I would judge the most delicious fruit would be a yellow raspberry direct from the cane on a warm day, or a William pear at the exact moment of perfection. But most of us have as much chance of these two experiences, or perhaps less chance, as of going one day to south-east Asia and eating a mangosteen in its perfection.

If such fruit, like others in this book, is to be savoured on its own, how should it be presented? Such things are not for casual eating. The dessert stage of a meal is, I suppose, the obvious occasion, the time for surprises and sheer pleasure when people are no longer hungry. Or it might go well in the late afternoon with wine, when friends call, and you can sit round a table talking.

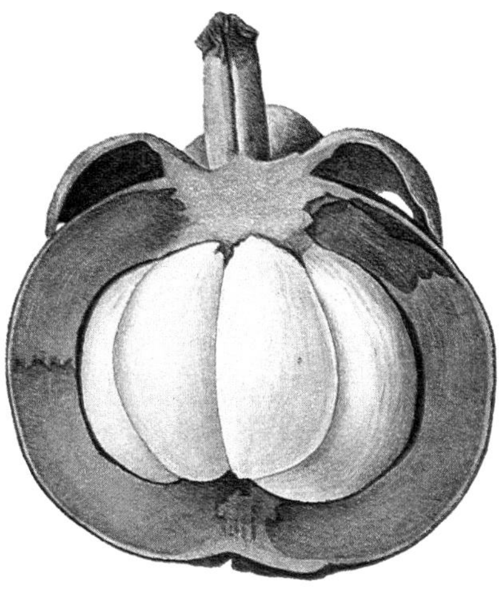

PASSION FRUIT

(Passiflora edulis)

The appearance of passion fruit does not come up either to their name or their fragrance. They are the most humble-looking objects, sometimes yellowish green and plump, but most often in our markets at any rate a dark, wrinkled, purple-brown. Cinderellas of the fruit world. Inside it's another matter. The acid, sweet-smelling pulp – full of edible black seeds, hence the name of grenadilla or granadill, meaning 'little pomegranate' – can be eaten on its own, or be used to flavour creams, ices and soufflés. Or it can serve as a lemon, a tropical sort of lemon, to bring out the flavour of other fruits. Which, I think, makes it unique among the exotic fruits of this book.

The name refers to the flowers, flowers that every gardener is familiar with, I would say, though from another variety, *Passiflora coerulea,* rather than *Passiflora edulis.* Their white, gold and lavender beauty caught the eye of the first Spanish missionaries in South America. In them they managed to decipher, from the intricacy of petals and stamens, all the signs of Christ's passion – the Three Nails, the Five Wounds, the Crown of Thorns and the Apostles (excluding Judas who betrayed him and Peter who denied him). They decided God had created this flower as an object lesson in the conversion of the pagans of this new continent.

I suppose they may have thought it appropriate that such a flower should bring forth fruit of such a sublime acid sweetness, though I have never read any comments to this effect.

When buying passion fruit, choose the least wrinkled ones. To eat them, simply slice off the top as if you were attacking a boiled egg. Dig in with a spoon, pouring in a little cream as you go. Passion-fruit fool is the logical extension of this simplicity, or passion-fruit jelly (see prickly pear, p. 35). The ingenious cook will soon become as used to having passion fruit as a standard ingredient, as to having lemons.

Passion-fruit Soup

5 passion fruit
melon weighing about 1 kg (2 lbs)
juice of a lime or lemon
150 g (5 oz) sugar
200 ml (7 oz) dry white wine

Halve the passion fruit and scoop out the pulp and pips. Cut the melon into chunks, discarding pips and skin; liquidize or blend it to smoothness, adding the passion fruit towards the end. Taste and use the citrus juice to bring out the flavour. Meanwhile dissolve the sugar in 500 ml (good ¾ pt) water over a low heat, then boil gently for 4 minutes. Cool and add to the fruit alternately with the wine, stopping when you have the balance that appeals to you. You can add cream or Greek yoghurt to this soup, but I like it best without. Chill, and serve as a first course, or as dessert with meringue kisses.

Kisses

Beat 3 egg whites and 120 g (4 oz) icing sugar together in a bowl over very hot water. Go slowly at first to mix in the sugar, then faster until you have a stiff, white, satiny cloud. The temperature should not be allowed to go higher than 50°C (120°F). Remove bowl, and continue beating until cool. Pipe into tiny meringues on to a metal sheet or tray lined with non-stick baking parchment. Bake them at gas mark ¼, 110°C (225°F) for about 2 hours. Check and be prepared to give them a little longer. Switch off the oven, open the door so that it just rests on the catch without closing, and leave several hours to finish drying out. Float a few on the soup, serve the rest in a bowl.

Passion-fruit Soufflé

pulp and seeds of 8 passion fruit
juice of a small lemon
level dsp gelatine
5 tbsps water
150 g (5 oz) sugar
3 egg whites, beaten stiff

Mix passion-fruit pulp with the lemon juice. Stir gelatine, water and sugar over a low heat until you have a clear liquid. Add passion fruit and chill until set to the consistency of egg white. Fold in the beaten egg whites and pour into a small soufflé dish, or 6–8 glasses. Chill to set.

Passion Fruit and Melon/Mango/Pawpaw Salad

Dissolve 125 g (4 oz) sugar in 125 ml (4 fl oz) water or white wine and bring to the boil. You can flavour the syrup with an orange liqueur or rum or citrus juice to set off the passion-fruit flavour. Off the heat stir in the pulp and seeds of 6 passion fruit. The heat, without cooking the fruit, will help to loosen the pulp round the seeds. By the time it is tepid to cool, you should be able to sieve it through. Keep back a few seeds. Cut up the quantity of melon, mango or pawpaw appropriate to the people you are feeding, and arrange it on a shallow dish. Pour over the passion-fruit syrup – there will be plenty for a salad for 6 – and scatter the reserved seeds on top. Serve cold but not over-chilled.

Passion-fruit Icing

If you hold back 3 or 4 tbsps of the passion-fruit syrup above, you can use it to flavour a cake icing. Sift icing sugar into a measure to give you ½ l (¾ pt). Cream 60 g (2 oz) soft unsalted butter with half the sugar. Add the passion-fruit syrup and enough of the remaining sugar to give you a smooth icing. Butter cream can also be flavoured with passion-fruit syrup or juice, so can a simple royal icing substituting passion-fruit juice for lemon or lime juice.

Maracudja

Fill a jar or empty rum bottle ⅓ to ½ full of passion-fruit pulp. Melt together 2 teacups of light soft-brown sugar and a teacup of water over a low heat. Bring to the boil and simmer for 2 minutes. Cool. Alternatively use light cane-sugar syrup. Measure out 1 part of the syrup you are using and add 4 parts of light rum. Pour on to the passion fruit until the jar or bottle is full. Close and leave for at least 6 weeks, longer, if possible. When you finally broach the jar, you can adjust the flavour to your taste by adding more rum or more sugar or syrup. Do not oversweeten. Serve well chilled as an aperitif.

African Passion-fruit and Melon Jam

Peel and cut up 2 kg (4 lbs) of melon, discarding the seeds. Put in a bowl with 1 kg (2 lbs) of sugar, cover and leave overnight. Butter a preserving pan and boil the melon and its liquid until clear. Scrape out the pulp of 36 passion fruit and add to the melon, with 8 tbsps lemon juice and 1 kg (2 lbs) sugar. Boil fast until setting point is reached. Bottle in the usual way.

PAWPAW

(Carica papaya)

In Mexican markets I was puzzled by stall after stall of vast objects, an orange-toned green, varying greatly in size. Some of them looked as if they might weigh 5 kg (11 lbs) or even more. When I asked about them, I was told they were pawpaws, and that pawpaws (often called papaya) do vary enormously, except for one or two varieties bred especially for commercial orchards – the solo from Hawaii and the papino from South Africa being the best-known examples. As their names suggest the fruit are small enough for one person to manage: they are the kind we see in our greengrocers and supermarkets.

The pawpaw is a giant plant rather than a tree. Its juicy green stem grows straight and tall, with a plume of leaves branching out from the top. When the fruit is formed it grows close under the leaves down the stem in a great cluster, the whole thing looking like an exotic Brussels sprout. James Grainger described it well in his poem on *Sugar Cane* (1764), writing of the 'quick pawpaw, whose top is necklaced round/With numerous rows of parti-coloured fruit.' The name pawpaw and its various pronunciations and spellings derive from the Carib word *ababai*: one gets more than a hint of this in the babaco fruit that are sometimes on sale in London – these are mountain pawpaws, and the fruit has to be cooked before you eat it.

The fruit makes a perfect breakfast in the tropics: you will be presented with beautifully cut apricot-pink slices and a wedge or two of lime – lime being the essential partner, as it is of a number of other mild-tasting tropical fruits.

Pawpaws can be used when green in soups, stews, chutneys and jams, when they begin to turn yellow (this is the stage when they are transported, since the skin becomes more tender as the fruit ripens and extra susceptible to damage) and when they are ripe. In general treat green pawpaws like squashes and marrows, and the pink-fleshed fruit like melon – with ginger, for instance, pepper, or in slices with thin slices of Parma, Westphalian, Bayonne or country hams, or with lettuce and in salads to go with smoked chicken or cooked ham or salted duck. Ripe pawpaw halves can be scooped out a little, filled with spicy beef and rice mixtures (include the flesh you remove, chopped) and baked in the oven; the cool, tempered sweetness sets off well the heat and zip of such dishes.

To Prepare

Halve the fruit and you will see that the cavity is full of deep grey seeds, the size of best caviare. They have a distinctive mustard taste, a greenish taste, and are not eaten. Scrape them out, then sprinkle the flesh with lime or lemon juice, if you are going to dig in with a spoon. Or slice and peel and season with lime juice, if this is appropriate, and a little sugar.

Pawpaw Chutney (South Africa)

1½ kg (3 lbs) prepared, diced green or partly ripe pawpaws
500 g (1 lb) green cooking apples, peeled, cored
500 g (1 lb) tomatoes, skinned
500 g (1 lb) seedless raisins
125 g (4 oz) peeled garlic
6 small hot chillies, seeded
500 g (1 lb) sugar
725 ml (1 pt 4 fl oz) wine or cider vinegar
1 level tbsp salt
small bag pickling spices

Mince fruits, raisins, garlic and chillies into a preserving pan. Add the remaining ingredients. Bring to the boil, stirring, then boil slowly until a chutney consistency is reached – up to 2 hours. Heap into hot sterilized jars (the mixture will contract as it cools) and cover next day. N.B. Never use metal tops with vinegar preserves.

Pawpaw Cheese (Nigeria)

For this you can use green, half-ripe or ripe peeled pawpaw, according to taste. Put the seeded, sliced and peeled pawpaw pieces into a pan and barely cover with orange juice or water. Simmer until soft. Blend, process or sieve the pulp and measure it in a measuring jug. Tip it back into the pan (rinsed out if necessary) and add an equal volume of sugar. Boil fairly fast, stirring, until very thick – the technique is the same as for damson cheese or apple butter. Make sure the fruit has no chance to catch. Brush out small, sterilized, straight-sided pots with glycerine (so that you can turn out the cheese easily to serve it). Cover immediately.

PERSIMMON

(Diospyros kaki)

A puzzling winter sight in Piedmont, in the yard of a suburban shack or out in a cottage garden in the country, is the brown skeleton of a tree, all leaves fallen, hung with huge orange-red globes. A glory often set off by snow and a blue sky. At first glance it seems as if someone had put up Christmas decorations a fortnight early, and that at night they would glow in the darkness. It's an artificial sight, and no wonder since in reality these are trees of Japanese persimmons, *Diospyros kaki*, kaki being the Japanese word for them and the name by which they are now known in much of Europe.

The great writer, Junichirō Tanizaki, describes their magical delight in *Arrowroot*, an account of a journey he made to the remote Yoshino district of west Japan. He was hunting down material for a novel about the semi-mythical Southern Court of the fourteenth and fifteenth centuries and was now sitting in the house of a family which still had some souvenirs of that distant time, poems, papers and a drum:

> As we were about to take our leave, the master of the house said, 'We don't have much to offer, but please try some ripers.' He made tea and brought out a bowl of persimmons with a clean, empty fire tray.
>
> 'Ripers' apparently means 'ripe persimmons'. The empty fire tray was to be used not as an ashtray, but as a dish from which to eat the soft, ripe persimmons. Following his urgings, I carefully put one of them on the palm of my hand. It looked as though it might burst at any moment. A large, conical persimmon with a pointed bottom, it had ripened to a deep, translucent red, and though swollen like a rubber bag, it was as beautiful as jade when held up to the light. The cask-sweetened persimmons that are sold in the city never achieve this fine color, however ripe they become, and they fall apart before they get this soft. Our host said that only thick-skinned Mino persimmons were suitable for making ripers. They are picked while still hard and sour and are put away where no breeze will strike them, in boxes or baskets. In ten days, without any human interference, the insides naturally turn to semiliquid as sweet as nectar. Other persimmons will get watery, not viscous like the Mino variety. By removing the stem and eating through the hole with a spoon, one can eat them like soft-boiled eggs; but they taste better if they are placed in a dish, peeled, and eaten by hand, though this method is rather messy. Just ten days will bring them to their most beautiful and delicious stage, but any longer than that will turn ripers into water.
>
> As I listened to this account, I gazed at the pearl of dew in my hand. It was as though the mystery and the sunshine of the mountains had congealed on my palm. I have heard that country people visiting the capital used to take packets of soil home with them as mementos; and if someone were to ask me about the color of

the autumn at Yoshino, I think I would take some of these persimmons home to show.

In the end, what impressed me most at the Ōtani house were the ripers, not the drum or the documents. Tsumura and I each devoured two of the sweet, syrupy persimmons, reveling in the penetrating coolness from our gums to our intestines. I filled my mouth with the Yoshino autumn. Even the mangoes of the Buddhist texts may not have tasted as good.

This perfect state is essential, for if you eat a persimmon too soon, while it is still firm and opaque, the tannic acid will set your whole mouth on edge. You may have to buy them hard, however. Put them into a bag with an apple – its ethylene gas which emanates naturally speeds the ripening process of the persimmons. Another way to ripen them is to put them into the deep freezer. The cold breaks down the tannic acid into sugars – presumably this is why they are left to hang on the trees in Liguria almost until Christmas. Once the persimmons are soft, you can cut off a lid and dig in with a spoon, pouring a little cream as you go. The pulp can be used for fruit mousses, ices, custards and creams, and even for a sweet-sour sauce to go with pheasant or venison, when a little sugar caramelized with wine vinegar can be used to give snap to the spiced mildness.

Greengrocers inevitably have problems with persimmons. Clever Israeli horticulturalists have found an answer to this with a persimmon that can be eaten when firm. It was developed from a Japanese variety in Hungary and is called Sharon fruit from Ha Sharon, where it was first grown on a large commercial scale. Sharon fruit lacks the coolness and glorious texture of a properly ripened persimmon, though it has its uses with smoked chicken, avocado and hazelnut in a salad.

The American persimmon, *Diospyros virginiana*, must also be eaten soft. It is now a rather neglected fruit in the States, as Raymond Sokolov describes in his book, *Fading Feast*, except in certain pockets of the South such as Brown County, where the fudge recipe below came from. These persimmons are reckoned – by Mr Sokolov at least, and I daresay he is right, since they grow wild and have not been 'improved' – to be far tastier than the Asian varieties, with date-like richness.

Brown County Persimmon Fudge

Persimmon pulp is made by putting the ripe pulp through a sieve or vegetable mill or by processing it, after scooping the flesh from the skin.

250 ml (8 fl oz) persimmon pulp
1½ kg (3 lbs) sugar
600 ml (1 pt) milk
125 ml (4 fl oz) light corn or golden syrup
125 g (4 oz) butter
125 g (4 oz) chopped nuts

Cook first 4 ingredients slowly together in a large heavy pan for 1½–2 hours, until the mixture reaches soft ball stage (230°F). Cool to lukewarm, stirring it up often. Beat in the butter and when the mixture starts to thicken as you continue beating, add the nuts. Pour into a buttered 20 × 30 cm (8 × 12 in) pan.

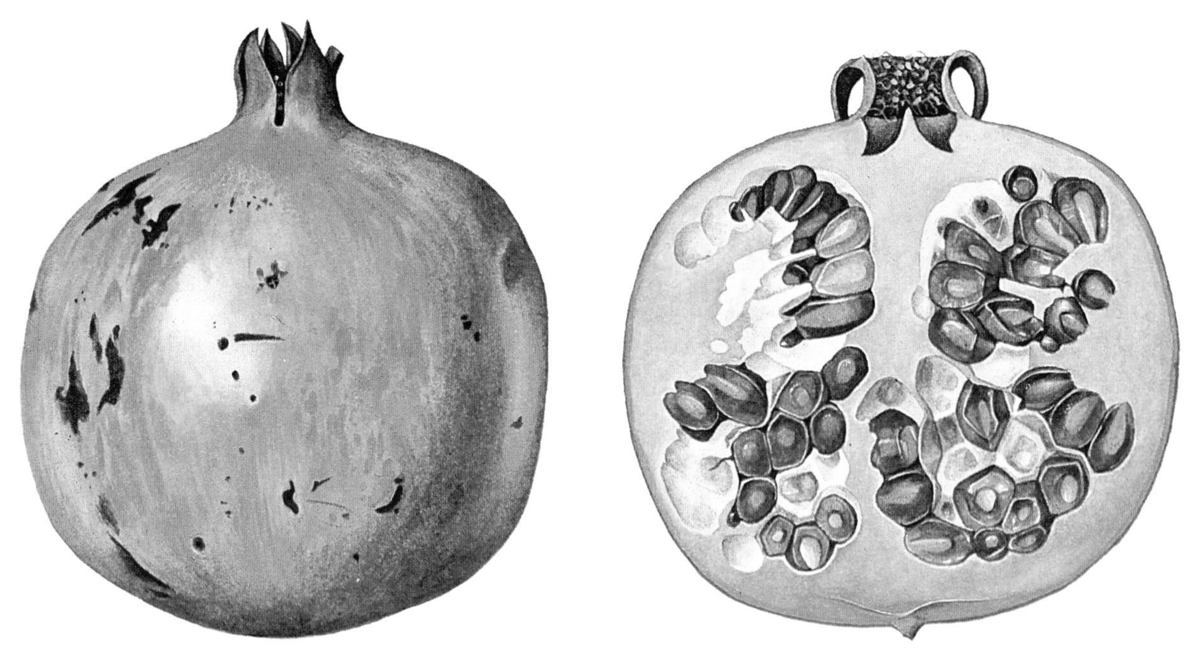

POMEGRANATE

(Punica granatum)

The Greek poet, George Seferis, talks of seeing pomegranates by the light of the setting sun. André Gide writes of their beehive partitions, an architecture of pentagons, skins splitting so that the seeds tumble out:

In cups of azure some seeds are blood;
On plates of enamelled bronze, others are
  drops of gold.

The seeds are the point – the name means grain apple and the Romans called it the Carthage apple, hence the botanical name of *Punica granatum* – each one enclosed in pinkish-purple translucent flesh, sharp, refreshing, grateful in the hot climates where the tree flourishes.

The tree, the red flower and the goldenred fruit are all beautiful in colour and form. As you would expect for a plant of such ancient enjoyment, there is a legend to account for the little crown that finishes off the fruit. Bacchus fell in love with a nymph who was of a cautious disposition, she had been told that one day she would wear a crown, and was saving herself for this royal prospect. Bacchus swore he would give her one. She believed him – and he turned her into a pomegranate with a calyx like a crown.

In spite of this mean little tale, the pomegranate was felt to be a noble fruit. Priests and kings had pomegranates embroidered in gold on garments of purple silk. Queens took the pomegranate as their device. The Moors planted a pomegranate avenue outside their city of Granada: in the city's crest was a pomegranate splitting open to show the grains. In Pluto's dark halls the only food Persephone was unable to resist was six pomegranate seeds, which is why we now have six months of winter. In Greek Orthodox rituals, dishes of wheat are decorated with pomegranate seeds, symbols of fertility, love, resurrection.

To Prepare

Halve and scoop out the seeds carefully, discarding the yellow pith; or peel the pomegranate like an orange, slicing off the crown, scoring down the rind and peeling it away in strips, to expose the seeds. If you just want the juice, halve the fruit and squeeze it in your hand, so that the juice runs through your fingers. Alternatively warm the fruit in a low oven, then roll and squeeze it gently in your hands to loosen the seeds. Finally make a hole in the stalk end and stand it in a glass to drip, giving an occasional squeeze of encouragement. If the juice tastes bitter, dissolve some gelatine in water and stir it in. This forms a cloud with the tannin, which can be trapped in a sieve.

Grey Mullet with Pomegranate Juice

In the neighbourhood of Comacchio in northern Italy – a place even more famous for its eels – marinated and grilled grey mullet are, according to Alan Davidson, served with a generous sprinkling of pomegranate juice instead of lemon.

Faisinjan

As in other rice-eating countries, Persians use meat, game and poultry cooked into thick stew-sauces as relish to the main, filling item (which means that the seasoning should be stronger than Western Europeans would consider appropriate). Faisinjan is the grandest of the Persian meat stews of this type.

Joint and season a duck or chicken, or cube a 1–1½kg (2–3lb) piece of boned lamb. Cook slowly in butter until browned and almost tender. Meanwhile soften a chopped onion in 60g (2oz) butter in a sauté pan, then thicken with 200–250g (6–8oz) shelled walnuts, skinned and ground. Pour in 200–250ml (6–8floz) pomegranate juice, the juice of half a lemon, plus about 750ml (1¼pts) beef stock. Stew for 15 minutes, uncovered. Add boned meat to the sauce. Simmer until tender. The liquid should be reduced, rich and thick by this time. Stir in the seeds of a pomegranate and adjust seasoning.

Serve with basmati rice, boiled or cooked in chilau style (soaked, boiled, then steamed with butter or oil so that the rice develops a crisp light-brown crust underneath).

Pomegranate Soup

Sweeten the juice of 10 pomegranates to taste. Slake 1 tbsp arrowroot with a little water. Heat the pomegranate juice till hot, but well below boiling point. Mix a little into the arrowroot, then return the mixture to the pan. Stir until clear and lightly thickened (arrowroot, unlike cornflour, does not need cooking for 2 minutes). Serve chilled as a dessert with cream and a few pomegranate seeds floating on top. Lemon juice may be needed to sharpen the flavour.

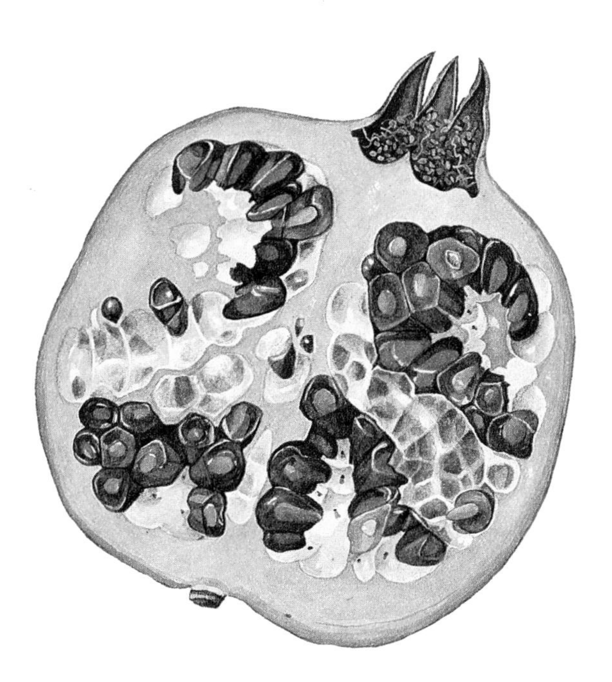

PRICKLY PEAR

(Opuntia ficus-indica)

Everyone knows, though they may not be aware of it, what the prickly pear cactus looks like. It is, I would say, our standard European image of a cactus, with its rounded sections branching out one from another in stiff spiky lines. By contrast few people know that you can eat the fruit, or that the second part of its botanical name means American Indian fig, as does the Italian *fica indiana*. The French call the fruit *figues de Barbarie* as if they had come from a ferocious stronghold of pirates on the Barbary coast.

The names pear and fig have, of course, no relationship to botanic reality. They refer vaguely to shape, as if travellers returned from the New World wanted to tell their friends the kind of fruit they had seen. 'Is it like a strawberry?' 'No, it's more pear-shaped, a bit like a fig perhaps.' The same thing happened with avocados. Fairly rapidly, after Columbus crossed the Atlantic and Spanish traders brought new things back to Europe, people had a chance to see such fruits for themselves. The prickly pear has travelled round the world, making itself at home in hot dry places. In some countries it is a ramping pest. In others it has its uses. Travel down to the south of Italy and you see plot after plot marked out with its barbed vegetation, as intimidating to animals as to man. In the country of Claude and Graeco-Roman ruins, it looks artificial yet magnificent, green paddles, yellow and red fruit, against a blue sky.

Outside Mexico, people do not make much of the young green cactus pieces that with tomato and coriander go into the *ensalada de nopalitos*. In Tijuana I was intrigued to see a market-seller rasping the thorny discs down a board which tore away the painful spikes, leaving a tender, harmless, juicy green. If you are curious you may find *nopalitos* canned in shops specializing in Latin American groceries.

The fruit travels well and in the north we begin to see prickly pears in the smarter supermarkets and greengrocers. They sell successfully, I suppose on account of Mediterranean holidays, when we see great piles of them, looking unfamiliar in markets or on roadside stalls. If you hesitate, someone will swiftly cut the skin open and hold it out so that you can pick the barrel-shaped fruit safely from it. It tastes sweet and bland, and cries out for lemon or lime. The colour goes from yellow to deep red, almost purple. The little seeds are perfectly agreeable. Not a fine fruit, not the best certainly, but welcome in the heat – and cheap, especially if you decide to pick your own.

To do this, split a garden cane down a little way. Push in a cork to keep the split open. With this in one hand and a stick in the other, you can dislodge the fruits from a safe distance. Wear rubber gloves, as it is easy to pick up a fruit that has tumbled to the ground, or fallen from your basket, without thinking. Then you have the devil of a job getting rid of invisible barbed hooks, pain and sudden anguish in your fingers.

To Prepare and Use

However smooth and prickle-free the fruits may look, do not be fooled. Handle them one by one with kitchen tongs, then pick up a sharp little knife, and a fork. Slice off each end, then cut – not too deeply – across the middle from end to end. Slip the knife in under the skin on one side of the fruit and gently ease it flat. Then repeat with the other side, until the barrel of juicy-looking flesh lies exposed.

For a salad, slice and sprinkle with sugar and lemon, lime or orange juice: then cover and leave for several hours. Serve cool rather than chilled. For a sauce, simmer in water to cover, and when tender, sieve, and flavour with citrus juice and sugar: serve with ices, creams, other fruit, adding an appropriate liqueur or spirit if you like.

Prickly-pear Jelly

Cook 6 prickly pears, sliced, with the pulp of 5 passion fruit in water to cover. Cook as little as possible – the aim is to loosen the seeds. Dissolve 1 pkt (½ oz) gelatine in 6 tbsps hot water. Add to the fruit in a measure. Top up with orange juice and water, to taste, until you reach 650 ml (1 pt 2 fl oz). Add a few black passion-fruit seeds, pour into glasses, chill until set and serve topped with layers of cream.

Prickly-pear Cheese

Cook sliced fruit with water to prevent sticking. Boil hard until the purée is jammy. Flavour with lemon and slivered almonds and pot in small jars brushed out with almond oil. Good with cream cheeses, junket, bread and butter, or to enclose in pastry and bake or deep-fry to make tiny kickshaws for the end of a meal.

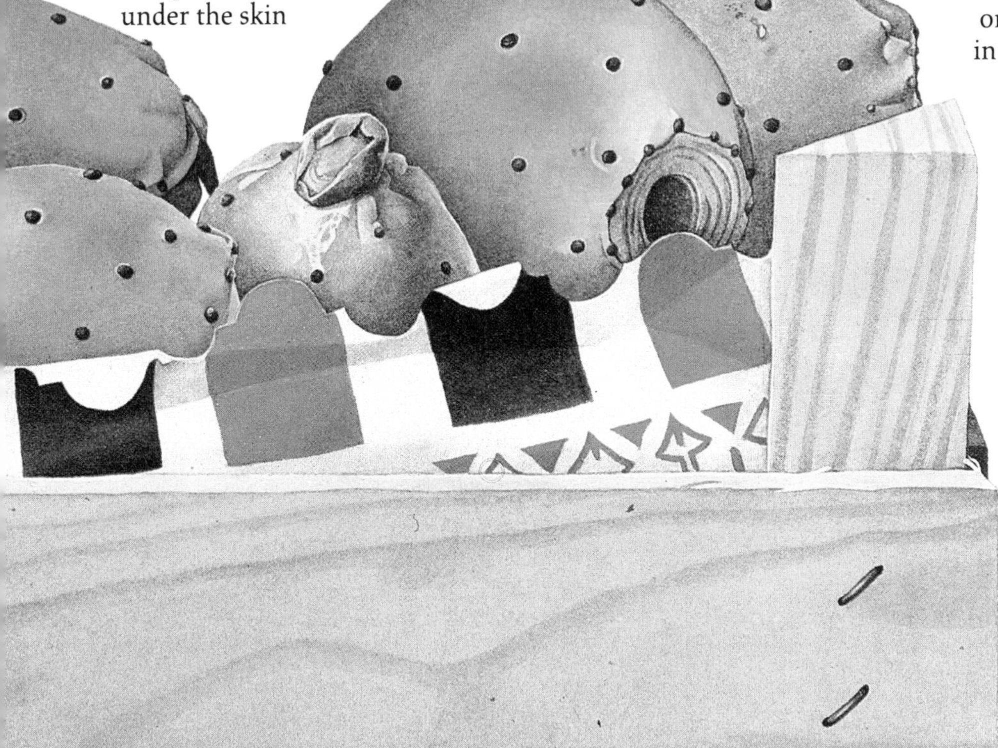

RAMBUTAN

(Nephelium lappaceum)

The most charming item of any large fruit bowl must surely be the rambutan from the Malay peninsula. This small, yellow and crimson to purple, unkempt hedgehog of a fruit, apparently bristling with spikes, is in fact soft to touch and easy to handle. The name means 'hairy one', *rambut* being Malay for hair. Sometimes it is called the hairy lychee, since the inner fruit has a similar greyish white translucency and texture. The taste, though, is not the same, but the rambutan can be used in the same ways – in sauces to go with duck, chicken or pork cooked in the sour-sweet manner, as well as in a *compote* or mixed fruit salad. But what a shame to do this, and deprive people of that monkey puzzle of an exterior.

An odd thing about the rambutan – and about the mangosteen, too – is that it has never conquered the tropical world. Think how the pawpaw and prickly pear from America flourish anywhere that the climate is right. Or the banana. Yet the elegant rambutan and mangosteen, both much appreciated on their home territory, have not spread far. Neither have they been taken up with any enthusiasm. When you think how long the British have been in Malaya, it is strange that these two fruits have not become a commonplace in this country. Surely colonial officials or retired officers and their wives living in Cheltenham must have longed for them? If so, they did nothing about it and for most of us without experience of the Malay archipelago the rambutan is as puzzlingly unfamiliar as the mangosteen.

For this reason take great care when buying them. Rambutans should not look dried up and brown and dead. Pick them over carefully, looking out for telltale softness, and choose the brightest fruit with the fleshiest spines. They should be tumbled rather than spiky, gentle on the fingers rather than dry. Colouring can vary, but I have a particular feeling for the deep crimson rambutan. They look intriguing and dramatic on a black dish. If you want to help your visitors, to give them a little information, cut a lid from the top of the shell which is leathery rather than crackling like a lychee's. Use small pointed scissors so that you do not damage the edible part inside. This gives you a third colour, a milky translucent white, to show off the crimson on black.

Occasionally you can buy lychees on the branch, several fruit connected by thin brown angular twigs, but I have never seen rambutan other than separately. A pity since they grow in clusters, sometimes of a dozen fruit, on trees that are stately and tall. I suppose this kind of consideration has no place in the realities of commerce, but as with clementines, which sometimes come with their leaves, such evidence of natural growth might make the fruit more attractive to people who hesitate before something so exotic and strange.

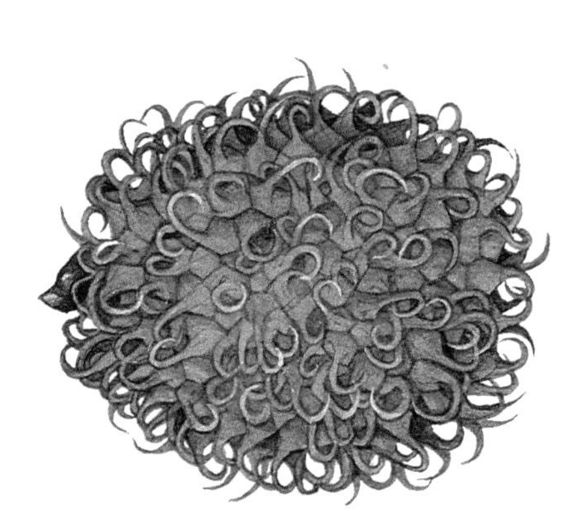

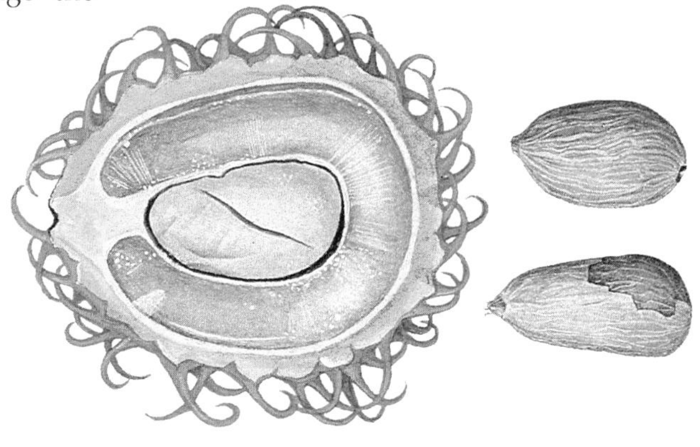

ROSELLE

(Hibiscus sabdariffa)

Roselle, rosella and rozelle are all names for a mallow originally from tropical Africa but at home in India for centuries (where it was prized for its fibre), in the West Indies and the Far East. These names are all corruptions of the French *l'oseille de Guinée*. I imagine that the red colour of the shrub, its red stalks and leaves, the red edible calyx, influenced a slide from l to r, from *l'oseille* to the more glowing sound of roselle. The plant's earliest name, its sixteenth-century name, was sabdariffa – nobody knows how or why or what it could mean – which was eventually incorporated into its botanic name.

The French name, meaning sorrel, applies rather to the taste than the colour of the fruit, which is not a fruit at all, but a boll or flower-cup of plump fleshiness which encloses the bud and seeds. It must be picked when young and tender. Then you can use it fresh, or dry it, as a kind of tropical cranberry.

Cooked up with some apple and water enough to cover, roselles make a sharp jelly for serving with poultry and game. They make a reasonable tart. Best of all they are turned into an essence or drink that comes in handy in hot weather. I notice lately that fruit waters are becoming popular again, as they were in the centuries before obnoxious 'squashes'. Partly because everyone has a refrigerator or freezer to store them in without the bother of sterilizing bottles, partly too because we begin to read the small print on labels of commercial fruit drinks and 'fresh' juices (made up from concentrate). When you have wondered at 'comminuted oranges', phenylalanine, phosphoric acid, aspartame, and felt uneasy at what such apparently innocent words as 'colours and flavourings' or 'artificial sweeteners' may conceal, the simplicities of redcurrant water – fruit, water, sugar – or *agua de Jamaica* made with roselles shine with honest truth and virtue.

Such things do not require recipes. Fill an enamel or glass saucepan about ⅓ full of roselles. Cover generously with water. Bring to simmering point, hold for 2 or 3 minutes. Remove from the heat, put on the lid and leave overnight. Strain and sweeten to taste. Use as a flavouring essence in drinks or to pour over ices; or chill and dilute with water – plain or fizzy – in tall glasses on hot days.

Christmas Sorrel Drink

Made in Trinidad for Christmas parties, since the sepals or flower-cup of the fruit are just right then. This recipe comes from Elisabeth Lambert Ortiz:

30 g (1 oz) dried sorrel sepals
7–8 cm (3 in) cinnamon stick
piece dried orange peel about 7½ × 2½ cm (3 × 1 in)
6 cloves
500 g (1 lb) sugar
6 tbsps medium dark Trinidad rum (or Appleton Estate or Mount Gay)
1 tsp ground cinnamon
¼ tsp ground cloves

Mix first 5 ingredients in a large jar and pour over them 1·8 l (3 pts) boiling water. Cool, then cover loosely and leave at room temperature for 2–3 days. Strain into a clean jar, add the rum and ground spices and leave for a further 2 days. Pour through a muslin-lined sieve and serve in chilled glasses, with ice cubes.

If you want a stronger cocktail, mix the drink with light Trinidad rum in the proportion of 2:1. Serve well chilled, with a couple of ice cubes and a squeeze of lime juice to each glass.

SAPODILLA

(Manilkara zapota)

Everyone, whether or not they realize it, has some acquaintance with the sapodilla, since the gummy juice tapped from its trunk is used to make chewing gum. This juice is known as chicle, from the Aztec *chickl*, their name for the tree.

Happily for those past childhood the sapodilla has another pleasure, its round or oval fruit encased in a green skin that turns brown as it ripens. This gives it a sad look, quite a mousy appearance, even more subdued than the kiwi's. And in export markets, in Britain anyway, it tastes as undramatic as it looks – or have I just been unlucky? You cut it through to a honey-fawn flesh, a softness that feels almost like a banana's, but with a taste of brown sugar. People have described it as pear-like, but I have found it softer. A squeeze or two of lime juice has a bracing effect on the floppiness, whether you eat the flesh as it is with a spoon, or mash it for use in a fool or a filling for meringues and other cakes.

Having said this, I should add that it would be imprudent to tackle a sapodilla that was not ripe, since the tannin in the early firm flesh can set your mouth on edge as it does with an unripe persimmon.

In fairness it does seem that nobody has made it his particular concern to ensure that exported sapodillas arrive on the market stalls of their destination in a state that is

reasonably close to their possible perfection. In the tropics – Central America for instance, the sapodilla's native home – people speak of the fruit with enthusiasm, even excitement. A Frenchman early in the nineteenth century, writing a flora of the Antilles, declared that 'an over-ripe sapodilla is melting and has the sweet perfumes of honey, jasmine and lily of the valley'. If you analyse this closely, it makes the sapodilla sound like an edible soap, but this was obviously not intended by the writer.

An incidental pleasure which no transportation, however bad, can take away is the long, svelte black seeds with a thin white stripe down one edge and a neat flick over at the top. Keep a few for their beauty in a dish.

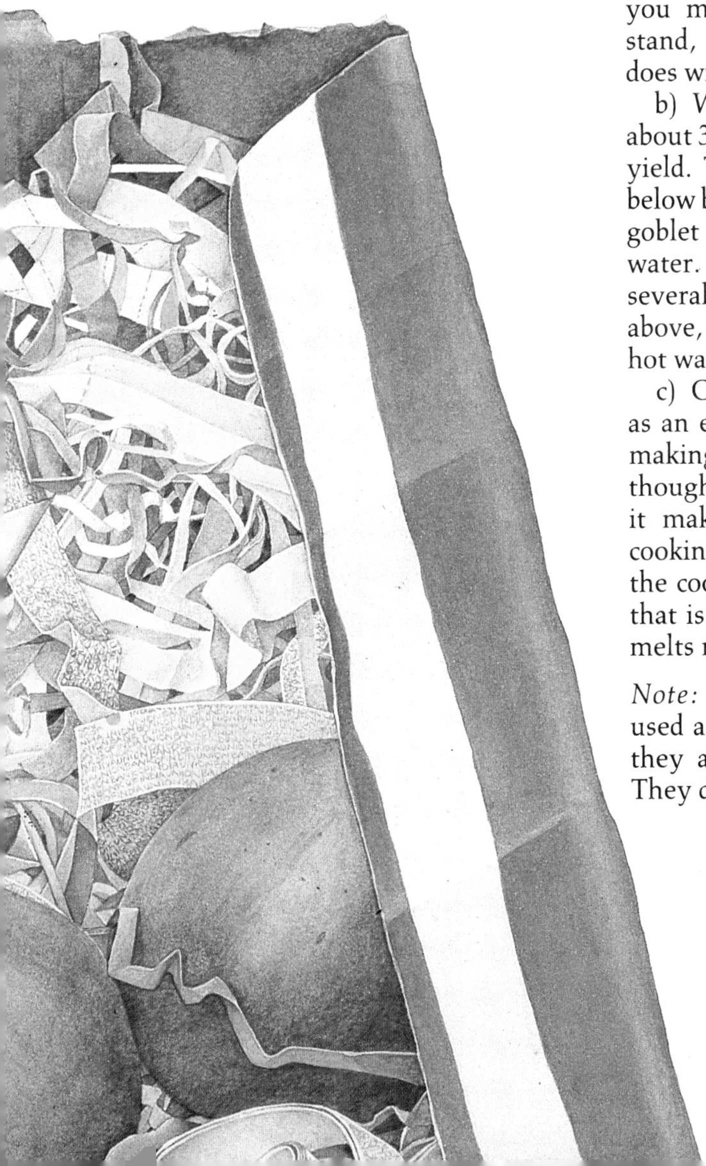

Coconut Cream and Milk

These are made from grated coconut, fresh or desiccated, or from blocks of creamed coconut dissolved in water. The liquid you hear when you shake a coconut is the water. Sometimes it is added to the coconut cream and milk, but it should never be confused with them.

a) Crack and remove the white part of a coconut. Grate it into a bowl and add a cup of water. Squeeze the two together until the water looks milky, then line a sieve with double muslin and pour the whole lot into it. Squeeze to get out all you can. This first squeezing is rich and thick, the cream in other words. Repeat the process twice and you should end up with approximately a generous half litre or pint of liquid. When you mix the two together and let them stand, the thicker cream will rise just as it does with cow's milk.

b) With desiccated coconut, you need about 375 g (12 oz) to equal a fresh coconut's yield. The water needs to be heated to just below boiling point. Put the coconut into the goblet of a blender and pour on the hot water. Leave 15–30 minutes, then whizz for several seconds. Squeeze through muslin as above, and repeat twice, each time with very hot water.

c) Creamed coconut in blocks can be kept as an emergency measure. Or use it when making a dressing for salad (as on p. 47), though the flavour is not so good. The cream it makes is unsatisfactory for prolonged cooking, but you can add a lump at the end of the cooking time to enrich a coconut sauce that is too thin in taste and consistency. It melts readily in heat.

Note: coconut cream and milk should be used and eaten by the end of the day after they are made. Store in the refrigerator. They do not freeze well.

VEGETABLES

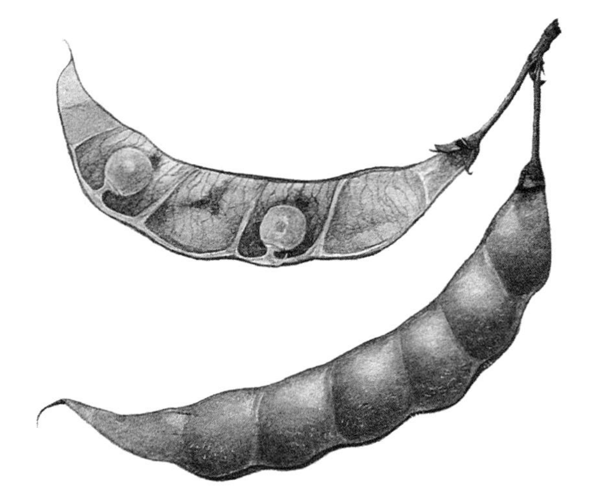

BEANS

Cowpea or Black-eyed Bean
(*Vigna unguiculata* subsp. unguiculata)

Although it is closely related to the yard-long bean, you are most likely to come across the black-eyed bean (*above left*) when it has been dried; use it as a pulse.

Pigeon Pea
(*Cajanus cajan*)

Should you find these peas (*below left*) in West Indian markets, they will be known as gungar. Discard the pod and try them in such dishes as rice and peas and pigeon-pea soup. They are dried in India and used in a type of *dal*.

Hyacinth Bean
(Lablab purpureus)

The hyacinth bean has a number of different cultivars, all varying in shape and size. Illustrated here are papri or popetti *(centre above)*, eat only the peas; seim *(centre below)*, with a thicker pod, is chopped diagonally and used in curries; valour or val *(above right)* looks very different but is topped and tailed and used in curries too.

Cluster Bean
(Cyamopsis tetragonoloba)

Cluster beans or guar (*below right*) are straight and pencil-like with a long, tapering point. Eat them as you would French beans or add them to vegetable curries. Commercially they are grown for their starchy gum, which is used in everything from ice cream to glue for postage stamps.

SA-TAW BEAN

(Parkia speciosa)

Sa-taw beans, native to the Malay archipelago, have a fairly restricted area of growth and use, but where they are eaten they are much appreciated for the characteristic flavour they give, lemony, a little sharp, slightly bitter, quite unlike any other legume.

At first, when the pods are tender and green, the whole thing is eaten. You slice it across, on the diagonal, and fry the pieces briskly so that the pod makes a good contrast to the soft bit of bean in the centre. Or the pieces can be added to a stir-fried dish of mixed vegetables, or poultry or fish and vegetables (see the following recipe). Later on, as the pods grow larger and undulate into curves and twists, they become too tough for eating. Not only do you need to shell the beans, you also need to cut away their outer skin, just as you do with broad beans if you really want to eat them at their best. In fact the sa-taw looks like a tropical version of the broad bean, although the beans themselves are more prominent inside the pods, not being as plumply cushioned. These beans can be bought in tins from oriental stores, packed in brine and labelled Peteh Asin – *peté* and *petai* as well as *peteh* are names from Indonesia, *sa-taw* is Chinese, *taw* or *tau* meaning bean.

In my experience you always need to shell and skin the beans if you manage to buy them in this country. The position may change, of course, so be on the look-out for the very young fresh kind. In any case buy them as green as possible, just as you would broad beans. As a start, try this lovely dish of Sri Owen's.

Sambal Goreng Udang (Prawns in Coconut Sauce)

1 kg (2 lbs) large prawns, peeled
100 g (3–4 oz) mange-tout or sugar peas
50 g (2 oz) fresh sa-taw beans, or drained peté asin
5 shallots, peeled, sliced
3 cloves garlic, skinned, sliced
5 macadamia nuts
1 slice trassi *or* blachan
5 fresh red chillies, seeded
1 tsp each ground ginger and ground coriander seed
pinch of powdered lemon grass
pinch of ground laos (galangal)
2 tbsps tamarind water (see p. 120)
2 Kaffir lime leaves or 1 bay leaf
1 tsp brown sugar (optional)
salt
½ cup thick coconut milk (see p. 39)
vegetable oil

Set the prawns aside for the moment. Top, tail and string the mange-touts, and skin and halve the sa-taw beans or halve the *peté*. Crush shallots, garlic, nuts, *trassi* or *blachan* and chillies to a paste, using a pestle and mortar. Add the ground ingredients to this paste. In a wok or sauté pan heat up a little oil and fry the paste, stirring it all the time, for 1–2 minutes. Put in the prawns and sa-taw beans or *peté*, then ¼ l (8 fl oz) water, the tamarind water, the Kaffir lime leaves or bay leaf, sugar and a little salt. Cook briskly for 6 minutes. Add the mange-touts and the coconut milk and bring back to something under the boil. Keep stirring for a further 2–3 minutes. Taste and correct seasoning. Serve immediately, with rice.

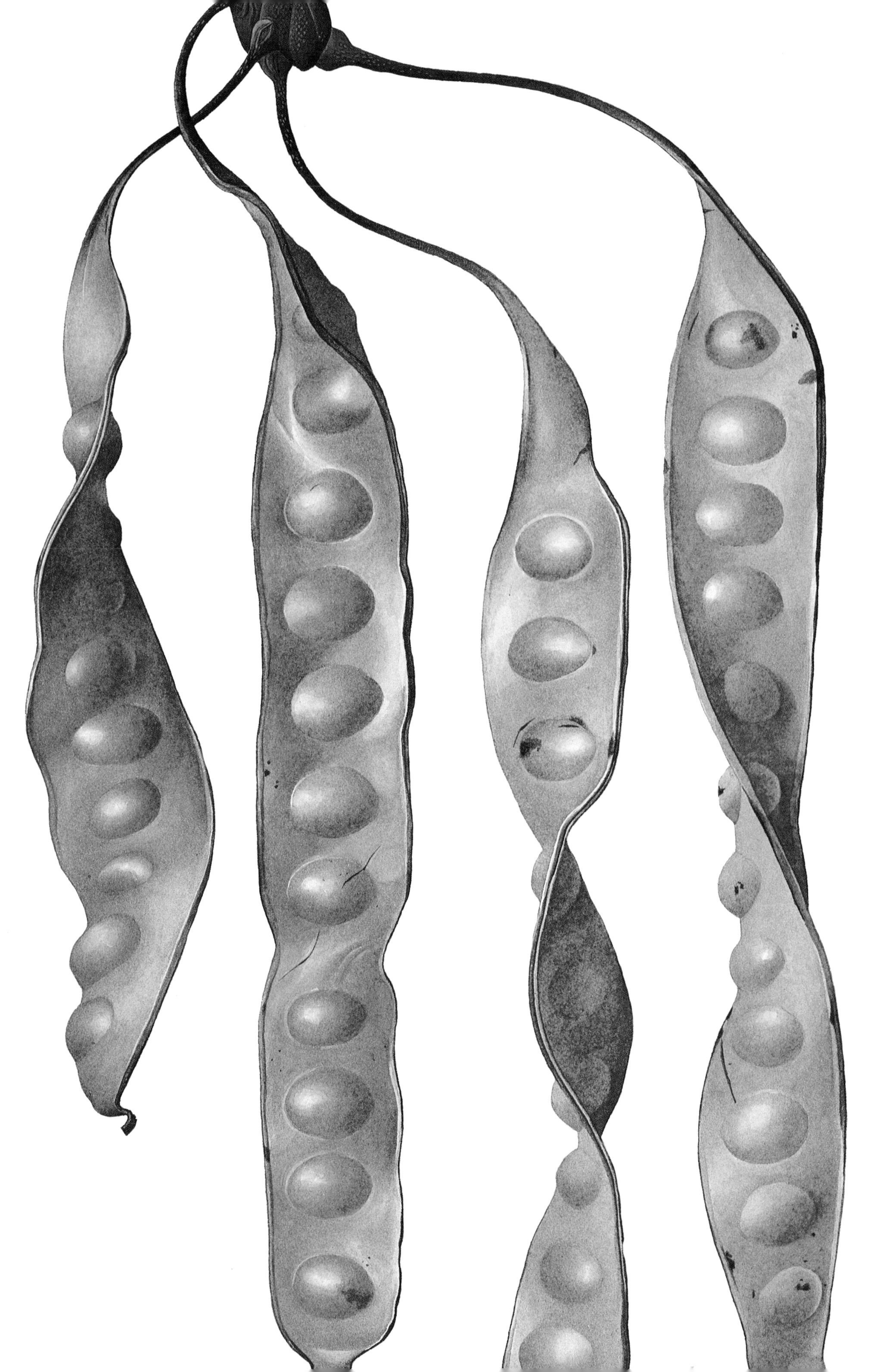

WINGED or ASPARAGUS BEAN

(Psophocarpus tetragonolobus)

The idea that a vegetable should grow frills down the whole length of its elegant greenness strikes me as bizarre. Yet there are two of them, not just one. The tiny asparagus pea of northern gardens, with miniature deep red flowers, and the winged bean, a much larger affair altogether, which also has beautiful flowers. So beautiful and decorative is the plant that in the Philippines they grow it all over their cottages, twining up and over and round every corner. It reminds me that when the scarlet runner was first brought to England by Charles I's gardener, John Tradescant, it was grown purely for its flowers and climbing habit. And judging by the size and toughness of the runner beans that some greengrocers have on display, it might be better if we once again regarded it primarily as a decoration for trellises.

The second part of the botanical name – *tetragonolobus* or four-lobed – refers to the four frills. What is misleading is the 'asparagus' part. There is no hint of asparagus in the bean's flavour or relationships, no hint in its shape and way of growing. The name is just a suggestion that we should eat

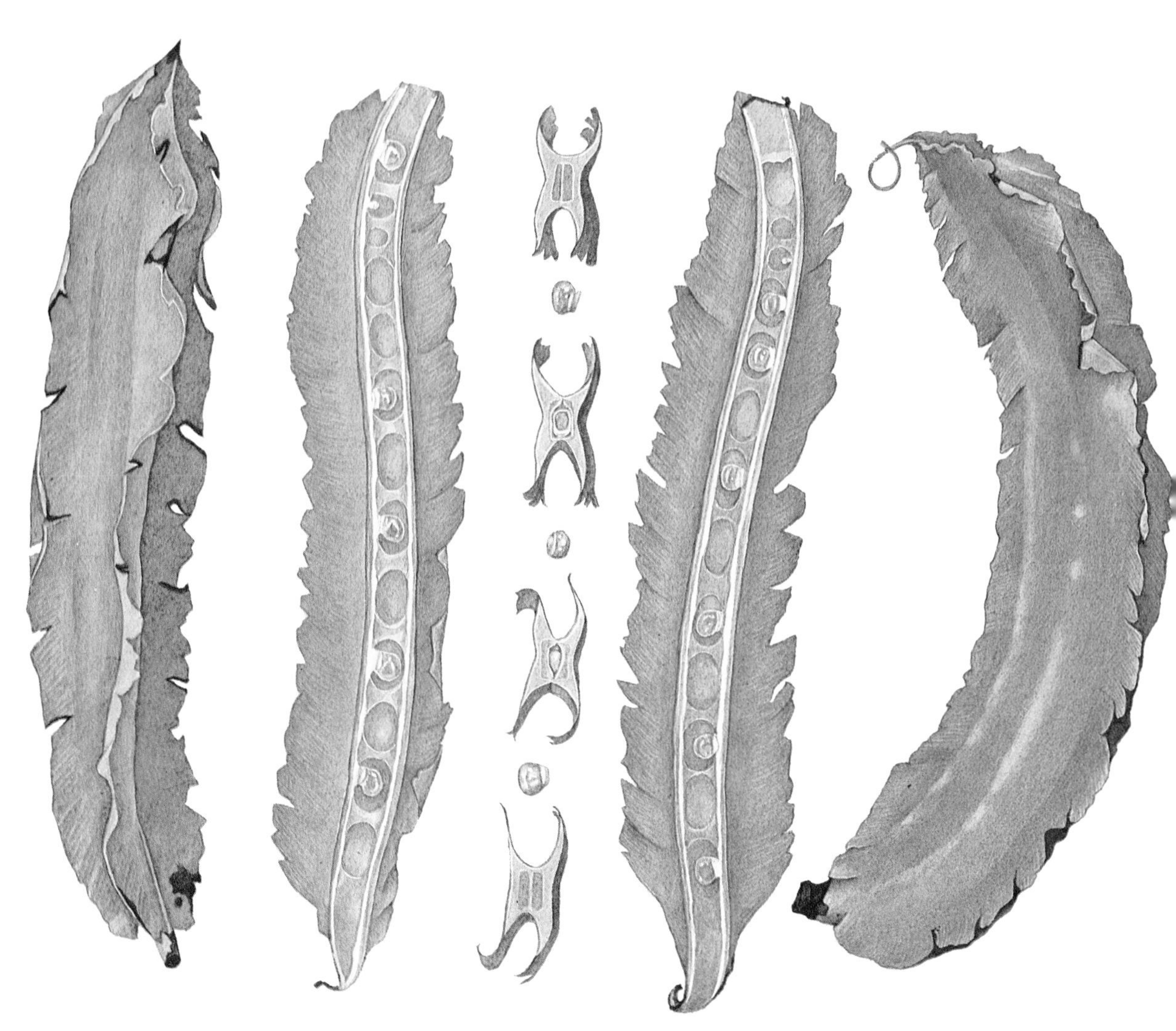

it, we Westerners that is, in the asparagus manner. Brief steaming or brief boiling, then picking it up in the fingers and dipping it in melted butter or appropriate sauce. The beans should be half-grown, about 10 cm (4 ins) long, for eating at their best.

In the Old World tropics – some scholars think it was a native of Madagascar – this asparagus or winged or four-angled or Goa bean is more likely to be cut up into salads and spiced dishes. The seeds inside the pod are also eaten separately and the tuberous roots, flowers and young shoots may be eaten raw as well as cooked; 'the flowers may be included in a salad to improve its appearance' – that is if you are a clever gardener and manage to get hold of some ripe seeds.

To Choose and Prepare

You are unlikely to have much choice, but go for the greenest, freshest looking ones, not longer than 10 cm (4 ins). Top and tail and treat generally like young French beans, unless you follow the asparagus style, or want to try out a dish like the following one (see recipes for yard-long beans as well). Reflect, too, that in English-language books of Far Eastern cookery, 'French beans' may well be a substitution for asparagus beans or yard-long beans which are not commonly available now, and were not available at all when some of the books were published.

Asparagus Bean Salad (Yam Tua Poo)

A salad from *Modern Thai Cooking*, by M. L. Taw Kritakara and M. R. Pimsai Amranand, part of a suggested meal for July. The rains begin, Buddhist monks study and teach in their monasteries, ardent young men make a retreat, and the women prepare food offerings. Among the dishes suggested are two containing asparagus beans. For the fish balls – *tod mun pla krai* – chopped asparagus beans or long beans are mixed with a freshwater fish and flavoured with a variety of seasonings that are not easy to find. Simpler to make, and good to eat, is the following asparagus bean salad. Measures are in cups, which is sensible for this kind of mixture – a 225 ml (8 fl oz) cream carton, and a couple more cut down to half and quarter cups work well if you do not have a set already.

1 cup sliced asparagus beans, cut thinly across
1 cup mixed boiled shelled prawns or large shrimp and cooked lean pork cut in similar slices to the beans
1–2 tbsps nam prik pow *(roasted chilli paste)*
juice of 1 lime
palm sugar or soft brown sugar
1 cup coconut cream (see p. 39)
salt
¼ cup sliced garlic
¼ cup sliced shallots
¼ cup dried chillies
vegetable oil
¼ cup crushed toasted peanuts
¼ cup grated coconut, lightly browned

Blanch the beans in boiling salted water until barely tender. Tip into a sieve then cool fast in a bowl of water chilled with ice cubes. They should remain a little crisp. Drain well and mix in a bowl with the shellfish and pork. This is the basic salad.

For the dressing, mash the *nam prik pow* with a little lime juice and ½ tsp sugar until it dissolves. Pour in ¾ of the coconut cream. Taste and add extra lime and sugar, plus salt.

Fry the garlic, then the shallots in a little oil until they are crisp but not burnt. Extract the seeds from the chillies, cut in shreds and fry them briefly, too. When cool, mix ⅔ of each into the salad with the flavoured coconut cream. Then mix in ⅔ of the peanuts and coconut.

Taste again. If the seasonings are too strong for you, soften the effect with the remaining coconut cream. Add extra salt, sugar and lime juice if you like.

Tidy the edges of the bowl if they are messy and scatter the remaining dry ingredients on top.

The recipes for yard-long beans (see p. 49) can also be used for asparagus beans.

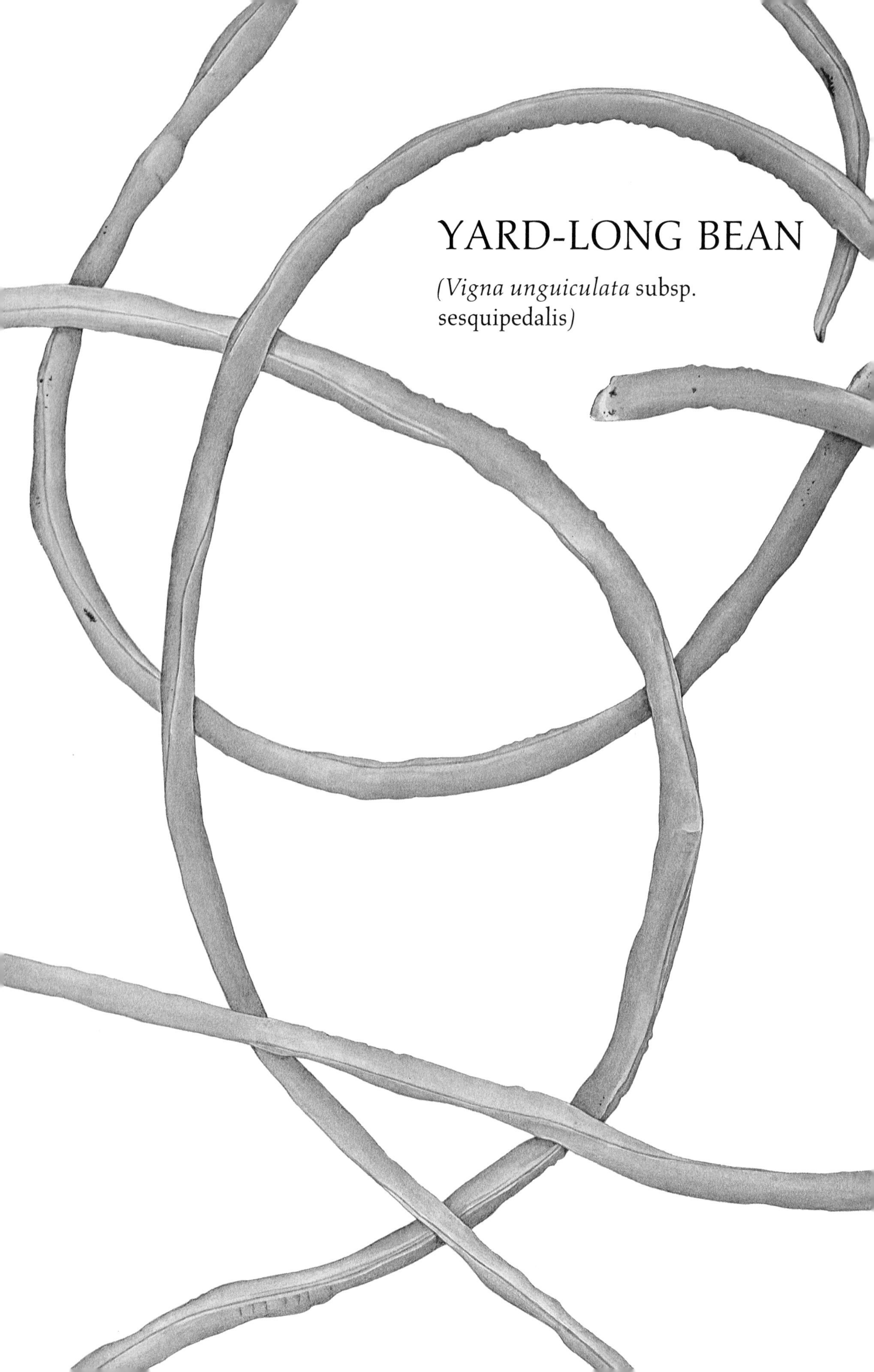

YARD-LONG BEAN

(Vigna unguiculata subsp.
sesquipedalis*)*

New vegetables that come in are often associated in my mind with particular people. Yard-long beans make me think of Thomas Jefferson, greatest of American Presidents, who never did quite manage to acquire seeds to grow them in his garden at Monticello. And of Sri Owen, who writes so nostalgically of the food of her native Java:

> The Javanese *warung* is the equivalent of the French pavement café . . . One of my two favourite *warungs*, all through my student days, was the *warung gado-gado* . . . The *warung* had a large table with benches on three sides and a nice old lady on the fourth. She chopped and cooked the vegetables, ground the spices for the *bumbu* or sauce, and whisked everything together at tremendous speed so that a dozen of us could all start eating more or less together.

To Choose and Prepare

For practical purposes yard-long beans are usually sliced into lengths. They can be cooked like French beans and served in a thoroughly West European style.

Stir-fried Pork with Yard-long Beans

In this simple, classic stir-fry, the important thing is not to overcook the beans, and to keep everything moving until the addition of the stock. Soak 1 tbsp fermented black beans in water for 30 minutes. Crush and chop a large clove of garlic to a mash, then add the drained beans. Heat 3 tbsps oil in a gwo or wok or commodious pan and fry the bean mixture briefly before putting in about 250 g (8 oz) cooked pork cut into dice. After 1 minute, put in 500 g (1 lb) yard-long beans cut into 1 cm (½ in) lengths, add a splash of soy sauce and cook for about 30 seconds. Pour in 175 ml (6 fl oz) chicken or pork stock. When everything is boiling, cover and lower heat so that the panful simmers. Test after 2 minutes to see if the beans are tender, then again after another 1. Finally mix in 1 level tbsp cornflour mixed to a paste with a little water. When the sauce is thickened, serve the dish.

Gado-gado

In Indonesia people do not pay the attention we do to eating hot food hot and cold food cold. If they did, this would be a very difficult salad to get right. Aim for a general warmth.

Sauce
125 g (4 oz) whole peanuts
oil
slice trassi
large clove garlic
2 shallots
salt
½ tsp chilli powder
⅓ tsp brown sugar
30 g (1 oz) creamed coconut
lemon juice

Vegetables
125 g (4 oz) yard-long beans, cut in lengths, boiled
125 g (4 oz) carrot sticks or cauliflower florets, boiled
150 g (5 oz) potato, boiled, sliced
red pepper, seeded, cut in strips, boiled
125 g (4 oz) cabbage, sliced, boiled
125 g (4 oz) bean sprouts, boiled
¼ cucumber, sliced or cut in sticks

Garnish
hardboiled egg, quartered
fried onion or chopped spring onion
prawn crackers (optional)

For the sauce, fry the skinned peanuts in 250 ml (8 fl oz) oil for 5–6 minutes. Tip them into a metal sieve over a basin and leave to cool. Pound or process them to a fine powder, or use an electric mill. Mash *trassi* and roughly chopped garlic and shallot to a nubbly paste with a little salt. Fry in 1 tbsp of oil for a minute, then stir in the chilli, sugar and a little more salt. After another minute add ½ l (18 fl oz) water. When it boils add the peanut powder and leave to simmer down to a thick sauce, stirring occasionally. Now add the coconut and stir until dissolved. Set aside until required. Arrange the warm vegetables on a dish. Reheat the sauce gently, add lemon juice and salt, if needed, to taste. Serve hot in a separate bowl. Add the garnish.

BREAD-FRUIT

(Artocarpus altilis)

People who know the story of the mutiny on the *Bounty* may never have realized – I certainly did not, until I began to take a special interest in food matters – that the ship had been sent expressly by the Government in London to take plants of the bread-fruit tree to the West Indies. Captain Bligh – 'Bread-fruit Bligh' – had been with Captain Cook on his second voyage in 1772–4, the voyage when they discovered bread-fruit at Otaheite. Bread-fruit was already known – Dampier in 1729 had described it as being 'as big as a Penny-loaf when wheat is at five shillings the bushel' – but it was on this second voyage of his that Cook thought it could be a beneficial addition to the slave diet in the West Indies.

The *Bounty* arrived at Otaheite in 1788 and spent five months there. This prolonged stay in the paradisal pleasures of Polynesia was balm to the crew. In his poem, *The Island or Christian and His Followers,* Byron uses the bread-tree as a symbol for the kind of existence that we picture from Gauguin's paintings:

The bread-tree which, without the
  plough-share yields
The unreap'd harvest of unfurrow'd fields,
And bakes its unadulterated loaves
Without a furnace in unpurchased groves,
And flings off famine from its fertile breast,
A priceless market for the gathering guest.

In fact bread-fruit is not such a perfect food as these lines might suggest. It needs to be eaten in huge quantities with coconut milk, fish and greens to give a balanced diet, because it contains so much water, as Dr Herklots points out: 'it remains a staple food in the Marquesas but elsewhere it is an item in the diet in which banana or taro is equally important'. Certainly the slaves in the West Indies did not appreciate bread-fruit so very much – a second ship, sent after the *Bounty* disaster, managed the trip successfully – and preferred the bananas they were used to. It has rightly been observed that people change kings, laws, and religions, more easily than they change their eating habits.

In Polynesia bread-fruit has been a staple food since prehistoric times when it was introduced from the Malayan archipelago.

> The open-boat journeys of the Polynesians in their peopling of the Pacific islands are marvellous from the point of view of seamanship alone . . . Probably a hundred species of plants were introduced into Hawaii by the Polynesians, and as a majority of their principal food-producing plants were propagated by cuttings alone, the difficulty in successfully carrying them across a wide expanse of ocean in open boats is obvious.

To Choose and Prepare

Take advice when buying a fresh, large bread-fruit. A nick from the stalk end should expose creamy to yellow flesh, rather than green, which means it is unripe. At home, ease and cut away the peel with a small sharp knife, keeping the surface as smooth as possible. It can now be cooked in various ways, like any starchy root vegetable: serve with plenty of butter, or some well-flavoured mixture (see p. 51). It is good to eat, with a genial hint of banana. Sometimes the seedy core is edible – in New Guinea they eat the seeds roasted or boiled and throw out the rest – but usually, or so I have found, it is tough and unrelenting.

Bread-fruit Crisps, or Chips

By analogy with game chips (and they do go well with game, or with a drink, or as part of a mixed first course). According to the quantity you require, slice a quartered, peeled and cored bread-fruit across into thin slices. Boil in batches for 2 minutes in salted water – a large pan of it, which does not go off the boil when you put a batch of slices in. Do not overload the pan. Drain and dry on sheets of kitchen paper. Deep-fry briefly until the slices look like potato crisps, nicely coloured and crunchy to eat. If you cannot stand the smell of deep-frying oil, remember that lard does well instead. Salt the crisps and keep them warm on baking sheets lined with kitchen paper, if you are cooking them in batches.

Baked Bread-fruit

Butter the peeled bread-fruit generously and wrap it in foil. Bake until tender – 45 minutes or more, according to size – at gas mark 4, 180°C (350°F). Cool slightly, then unwrap and remove the core. Serve with plenty of butter, salt and pepper. Or with a good yoghurt, e.g. Greek Total cow's milk yoghurt, rather than butter. Or stuff it with a piquant ragout of salt cod, which turns the whole thing into a main course dish.

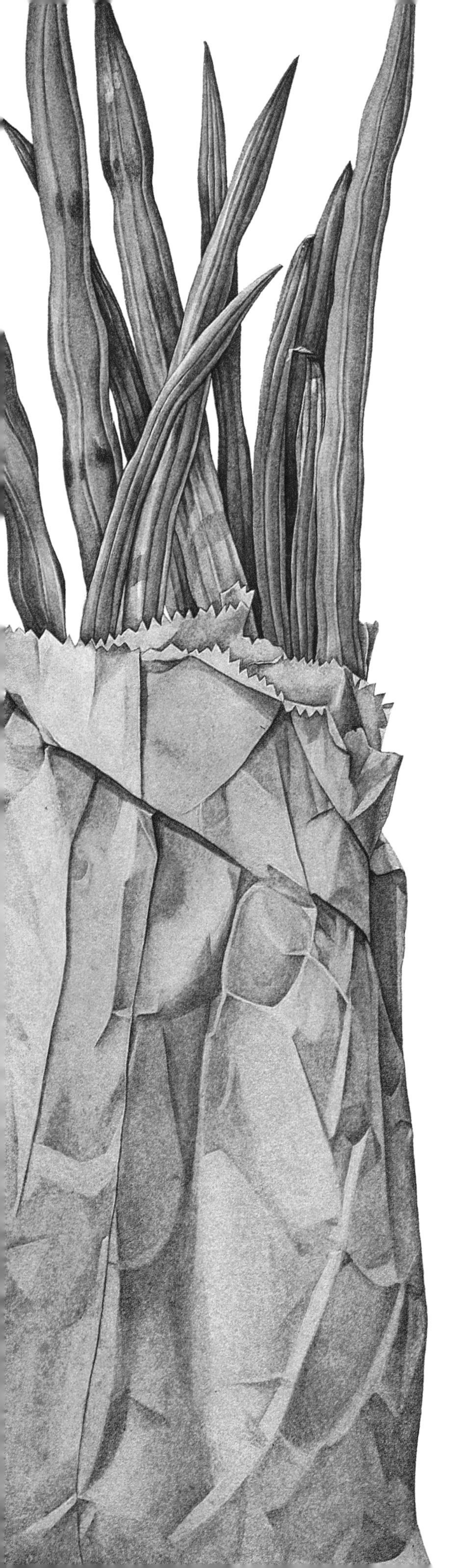

DRUMSTICK

(Moringa oleifera)

The elegant beauty of the drumstick, long and green with ridges, is most seductive, but I confess to finding the cooked reality disappointing. Perhaps the examples I have tried were not as young and tender, as newly picked, as they should have been, because it is a vegetable that has been eaten with pleasure by Europeans living in India. Drumsticks (the name was given by the English) are the fruit of the horse-radish tree – horse-radish, another English name, because that is what the grated root tastes like. I have this picture of overclad families, far from home, the punkahs going, drops of sweat discreetly mopped away, plodding their way through roast beef and Yorkshire pudding with the aid of mock horse-radish sauce – 'Not bad at all, my dear!'

To Prepare

Scrape away the toughest parts of the skin, then rinse them and cut into 8–10 cm (3–4 in) pieces. Simmer them in salted water until almost cooked.

Drumstick Foogath

250–375 g (8–12 oz) drumsticks
4 peeled slices fresh ginger
3 large cloves garlic, skinned, crushed
1 tsp turmeric
¼ tsp cumin seeds
½ tsp hot chilli powder, or 1 fresh chilli, seeded, crushed
½ tsp salt
ghee or clarified butter
125 g (4 oz) sliced onion
2 tbsps grated coconut, fresh if possible, otherwise desiccated

Prepare drumsticks as above. Crush together or whirl in an electric mill the ginger, garlic, spices and salt.

Heat enough ghee or clarified butter to coat the base of a pan that will take the drumsticks more or less in a single layer. In it cook the onion to a lightly browned golden stage. Stir in the paste (i.e. the garam masala) and continue stirring for 1½ minutes.

Add the drained drumsticks and stir-fry gently – so that the pieces do not break up and lose their definition – then scatter in the coconut and continue to cook until the drumsticks are cooked. Taste and add extra salt if necessary.

Drumstick Curry with Prawns/Shrimp

250 g (8 oz) drumsticks
500 g (1 lb) uncooked prawns or large shrimp, or 375 g (12 oz), if cooked
60 g (2 oz) chopped onion
1 tsp ground chillies, or 1 small fresh hot chilli, seeded, crushed
1 tsp paprika
½ tsp ground turmeric
1 clove garlic, crushed
1½ tsps salt
mustard or sunflower oil
1 tsp tomato concentrate (optional)

Prepare the drumsticks as above. If the prawns or shrimp are uncooked, give them a few minutes in 300 ml (10 fl oz) water; cool them, in a sieve, retaining the liquor. Then shell and devein them. If the prawns or shrimp are already cooked, shell them and simmer the shells in 300 ml (10 fl oz) water to make a little stock. By either method, aim to end up with 250 ml (8 fl oz) stock.

Crush together the onion, spices, garlic and salt to make a paste. Coat the base of a large pan with mustard or sunflower oil. Fry the paste for 1½ minutes. Stir in the drumsticks and prawns or shrimp with the liquid. Cook gently until tender. Taste for seasoning towards the end of cooking time and add the tomato concentrate gradually if it seems a good idea – it intensifies the flavour, without being recognizable.

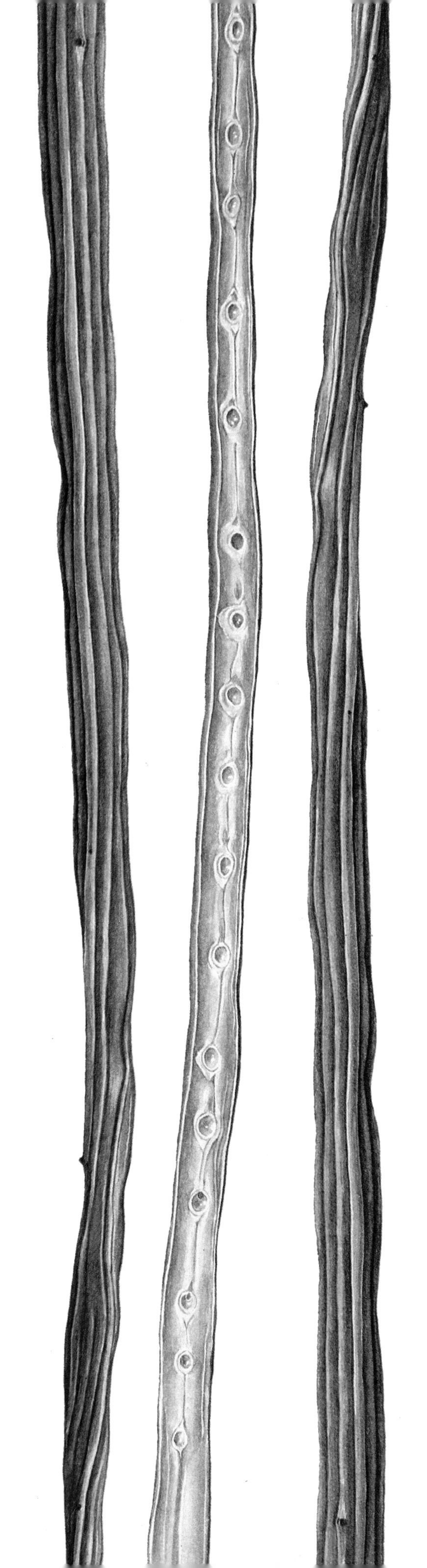

Angled loofah or turia
(*Luffa acutangula*)

GOURDS

ANGLED LOOFAH and SMOOTH LOOFAH

(Luffa acutangula and *Luffa cylindrica)*

When I was a child we lived in a tall Edwardian house that had been very well and idiosyncratically fitted out for its original owner. He had also owned a shipyard at a time when such places flourished. No doubt this accounted for the quantity of well-aged timber and skilful carpentry. The doors swung shut with a quiet, rich click, the cellar had a sprung dance floor (ideal for roller skating, too). We did not really live up to that house.

The bathroom in particular was awe-inspiring. Not only did it have an immense bath which turned yellow when the wind blew sulphur fumes over from the pit villages, but the bath itself had a tall, curving end-screen fitted with many vertical pipes that shot boiling or icy water at you in minute prickly jets. And it contained two strange objects of indefinable substance. They did not float or do anything amusing. I remember being told that they were loofahs to wash your back with – and then being given a vigorous demonstration which made me nervous of them ever after. Even then it seemed to me that the explanation was inadequate. I had seen sponges in other people's houses, lovely soft things: did loofahs grow under the sea, too? That stumped everyone.

Now I know at last that the loofah is a gourd, or rather the fibrous internal structure of a particular gourd: what is left when the green outer skin has been pulled away from the dried-out gourd, and the rest has been soaked and retted for several days in water so that the soft part and seeds can be pulled away leaving the skeleton of a structure. This is then bleached and flattened, ready for use. The smooth loofah, *Luffa cylindrica*, once known as *Luffa aegyptica*, is the one usually treated in this way. The angled loofah has the same structure but it is much trickier to remove its skin and flesh, so for commercial purposes the smooth kind is preferred. These days the great country for loofahs is not Egypt but Japan, where they produce 24,000 per acre. They can be used for filters, for cleaning forks (you thrust the prongs in and out several times), for stuffing tablemats and saddles, for the soles of slippers. I suspect that they became popular in bathrooms here with the arrival of Turkish baths in the middle of the last century. David Urquhart, diplomat and lover of Turkey, praised them in his book, *The Pillars of Hercules* (1850). Soon afterwards he superintended the building of the Turkish baths in Jermyn Street. In the final stages of what became a smart and sociable activity people were rubbed down with loofahs.

Both smooth and angled loofahs should be eaten when young. Later on they develop a bitter flavour. Philip Miller, the great gardener of the Chelsea Physick Garden, was the one who gave the smooth cylindrical loofah its first botanical name of *Luffa aegyptica* (loofah comes from the Arabic *luff*, hence the botanical name of the genus; this species, one of seven, was first spotted growing in Egyptian gardens in the seventeenth century). He commented on the elegantly netted fibrous inside, but concluded, 'The fruit, when it is young, is by some people eaten, and made into Mangoes, and preserved in Pickle; but it hath a very disagreeable taste, and is not accounted very wholesome; wherefore it is seldom cultivated in Europe, except by such persons as are curious in Botany, for variety.'

John Organ, author of *Gourds*, concurs with this opinion and says that although one may become accustomed to the flavour, as one does to retsina, no European can ever thoroughly enjoy it. Certainly man did not rush to take the loofah into his garden and kitchen – or into his washing habits, come to that. Both varieties are natives of the Old World tropics, probably of India, but there is no name for them in Sanskrit and no record of their early use in China, although nowadays they are part of the extraordinary range of vegetables grown in southern China (in Hong Kong markets you will find very small angled loofahs which are so tender they do not need peeling). In the Caribbean, where squashes are legion in their variety, they are treated much like cucumber or chayote – simply boiled and served with butter and plenty of pepper, or else stuffed with some meat or fish mixture and then baked.

Other names for this loofah emphasize the homely nature of its usefulness: dishcloth gourd, vegetable sponge, towel gourd, African sponge.

To Choose and Prepare

Pick out the smallest, greenest loofahs, since the larger they grow the more likely they are to be bitter. In any case they should never be sold if they are longer than 20 cm (8 ins). Peel them thinly and slice or cut them into strips according to the recipe. The Chinese stir-fry them with pork and chicken, or in general vegetable mixtures. You can cook them like cucumber, courgettes and so on. Another way is to scoop out the seedy inside, fill it with stuffing and then cook it.

Petola Daging

In other words stuffed smooth loofah or bottle gourd in the Sulawesi style – and dialect – from Sri Owen's *Indonesian Food and Cookery*. Many other gourd-style vegetables can be cooked in the same way, and of all the meat and fish stuffings I would recommend this one for the lightness of the stuffing, which comes from mixing omelette strips into the other ingredients. Much better than a pounded smoothness. Try it with bitter gourd (karela) or cucumber, after parboiling them: if you intend to bake rather than steam the loofahs, it is a good idea to parboil the pieces, too, to give them a start.

5 smooth loofahs, thinly peeled
125 g (4 oz) prawns, shelled
250 g (8 oz) minced lean beef or chicken breast
4 eggs
salt, pepper
vegetable oil
5 shallots or 1 onion, thinly sliced
2 green chillies, seeded, sliced, or ½ tsp chilli powder

Halve the loofahs longways and scrape away the seeds. Leave in cold salted water until the stuffing is completed.

Chop the prawns and mix with the meat. Separate 1 of the eggs, keeping the white to one side, and mix the yolk with the remaining 3 eggs; add seasoning and cook several thin omelettes. Let them cool, then roll them up and cut across into thin slices that look like pieces of tiny Swiss roll.

Meanwhile heat a little oil in a wok or sauté pan and cook the shallot or onion and chilli gently until soft. Do not brown them. Add the prawn mixture and stir-fry for 3 minutes. Season to taste, allow to cool slightly and then stir in the single egg white. Mix in the coiled omelette strips.

Drain and dry the loofah halves. Put the stuffing into 5 halves and put the remaining halves on top like lids. Steam for 30 minutes, or bake for 40 minutes in the oven preheated to gas mark 4, 180°C (350°F), in a lightly oiled dish. Eat hot or cold, sliced thickly across.

Toorai Tamat (Loofah in Tomato Sauce)

1 kg (2 lbs) loofahs, peeled
2–3 tbsps ghee or sunflower oil
1 large sliced onion
2 large cloves garlic, finely sliced
1 cm (¼ in) piece ginger root, peeled, chopped small
250 g (8 oz) tomatoes, skinned, seeded, chopped

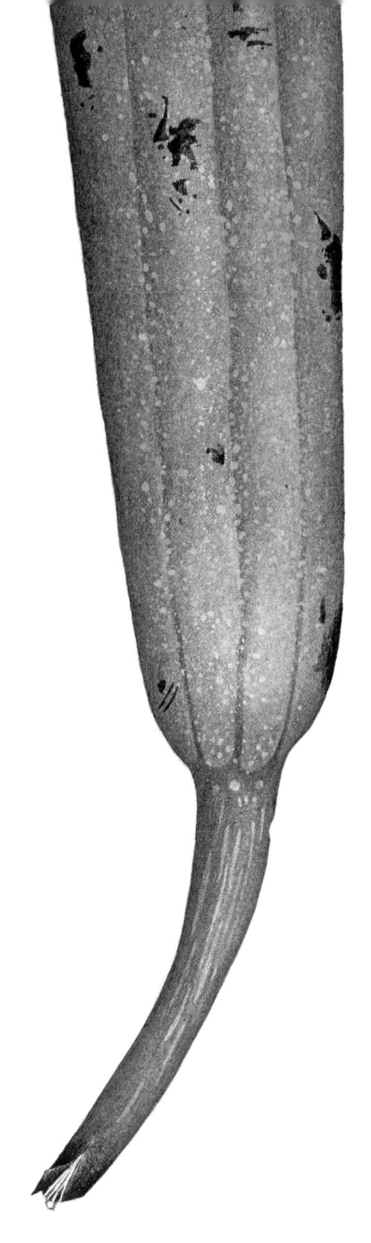

Smooth loofah
(*Luffa cylindrica*)

1 large chopped onion
1 tsp salt
1 tsp sugar
tomato concentrate or purée (see recipe)

Slice the loofahs thinly. Heat enough ghee or oil to cover the base of a large heavy pan and stew the sliced onion in it until golden and tender, not brown. Pour off any surplus fat. Put in garlic and ginger and give them 3 minutes gentle cooking. Add tomatoes, loofah, the chopped onion, salt and enough water to come about 3 cm (good inch) up the pan. Bring to boiling point, then leave uncovered until the loofah is tender and the water almost completely gone, leaving a general juiciness. Add sugar, cook for a minute or two, stirring, and taste. The tomato flavour may need reinforcing with concentrate or purée as northern tomatoes are so tasteless.

Serve on its own, with some bread for instance, or with samosas or other little pastries, with chops and with dishes of minced meat.

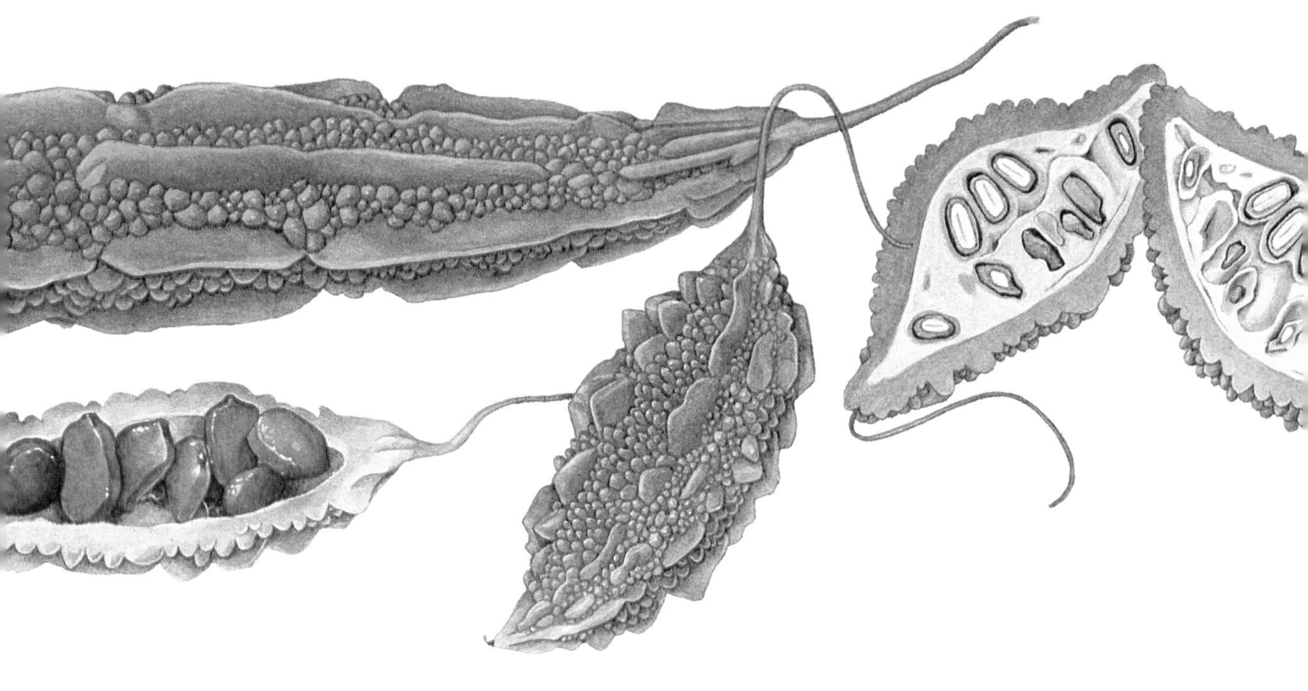

BITTER GOURD and SPINY BITTER GOURD (KARELA and KANTOLA)

(Momordica charantia and Momordica cochinchinensis)

These two are the most commonly eaten of the forty-two species of the *Momordica* genus, all from the Old World tropics not America, which is the country most of us think of when some new strange gourd comes into the shops. In fact we have known about these two since the sixteenth century but they have conquered neither our tastes nor our conservatories and greenhouses (where they can be grown successfully). Why should this be so?

The most vivid answer to that question is Rudyard Kipling's song from the *Second Jungle Book*, or rather 'Mowgli's Song Against People':

I will let loose against you the fleet-footed
 vines –
I will call in the Jungle to stamp out your
 lines!
 The roofs shall fade before it,
 The house-beams shall fall;
 And the *Karela*, the bitter *Karela*,
 Shall cover it all!

In the gates of these your councils my
 people shall sing.
In the doors of these your garners the
 Bat-folk shall cling;
 And the snake shall be your watchman,
 By a hearthstone unswept;
 For the *Karela*, the bitter *Karela*,
 Shall fruit where ye slept!

In other words the answer is bitterness, what one writer calls a quinine bitterness, something we have not – in the past at any rate – much appreciated in our diet. Kipling, too, sees the bitterness as a kind of vengeance, not as a stimulus to appetite. This must surely be a very Western point of view. Would an Indian be anything but delighted if his house were suddenly covered with bitter gourd – or, come to that, with spiny bitter gourd, although it is not as bitter? Would he see the 'bitter herbs' of the Bible for instance as a punishment? Perhaps with our new taste for chicory and endive, we shall come eventually to these two vege-

Bitter gourd or karela
(*Momordica charantia*)

tables that are such delicacies in the Far East, in China as well as in India and Pakistan. At the moment you need to go to West Indian and Indian stores and markets for them.

To Choose and Prepare

These gourds are eaten when they are very young, and the bitterness is enjoyable, so choose the smallest. Scrape them to remove the coarsest part of the skin. Then, according to the recipe, slice the gourd across or make a slit in it, so that the inner seeds and soft pith round them can be removed. They can then be salted – 2 tsps salt sprinkled over 450 g (1 lb) bought weight. After 3 hours, the salt and much of the bitterness that has emerged in the juice can be washed away. Another alternative, especially when the gourd is young (taste a tiny bit to see), is to boil it for 3 minutes, then drain it well.

Both the bitter gourd and the spiny bitter gourd can be used in the same ways, although the spiny bitter gourd may need less time salted. Ideally you should buy them very young, about 2 cm (less than 1 in) long, when they can be cooked whole, skin, seed and all. They can go into curried dishes with a number of ingredients, or they can be stir-fried or stewed as a vegetable on their own.

Stuffed Bitter Gourd

Scrape and salt 6 large bitter gourds. Slice each across to make 4 deep rings and clean out the centres. Weigh them. Process half their weight in whiting fillet and half their weight in shelled prawns or shrimp, together with a small egg white for every ½ kg (1 lb) of fish and shellfish together. Season. Fill the rings, mounding the filling at each end. Fry in a little oil on the filled ends. Remove them from the pan.

Meanwhile for every ½ kg (1 lb) of gourds, soak a slightly rounded tsp of fermented black beans for 10 minutes. Rinse them and mix with a small finely chopped clove of garlic. Fry lightly in the gourd pan. Pour in 250 ml (8 oz) chicken stock and ¼ tsp sugar. Stir up well. Put in the gourd pieces, cover and leave to simmer. After 10 minutes test them, and be prepared to give another 10 or even 15 minutes if the gourd is thick: remove the pieces to a bowl. Slake 2 tsps cornflour with 1 good tbsp stock and add to the sauce. Bubble a moment or two to thicken and clear it. Taste for seasoning and pour over the pieces.

Bitter Gourd with Tomato Indian Style

Serve as a dish on its own with rice or Indian bread, yoghurt (perhaps in the form of lassi, i.e. as a drink, liquidized with an equal volume of water, with a little salt, plus an ice cube or two in the glass) and a *dal*; or as a vegetable in the Western way with chicken, duck or goose.

8–9 medium bitter gourds, prepared, sliced, salted
1 tsp slightly crushed coriander seeds
1 tsp fennel seeds
3 tbsps oil
300–375 g (10–12 oz) chopped onion
250 g (8 oz) tomatoes, skinned, chopped
175 g (6 oz) drained canned tomatoes, chopped
2 cloves garlic, crushed
2 hot green chillies, seeded, chopped
salt
½ tsp ground turmeric

Blanch the drained and rinsed gourd slices for 3 minutes in boiling salted water if they are very bitter even after the preliminary salting. Pour off the water and set aside.

Fry coriander and fennel in the oil until it pops, a matter of seconds, then stir in onion and cook until it is soft and yellowish. Put in both lots of tomato, garlic, half the chilli, a scant tsp salt and the turmeric. When you have a pulpy sauce, add the gourd slices and simmer until they are tender but still slightly crisp. Taste occasionally, adding extra chilli to taste. Stir often.

Spiny bitter gourd or kantola (*Momordica cochinchinensis*)

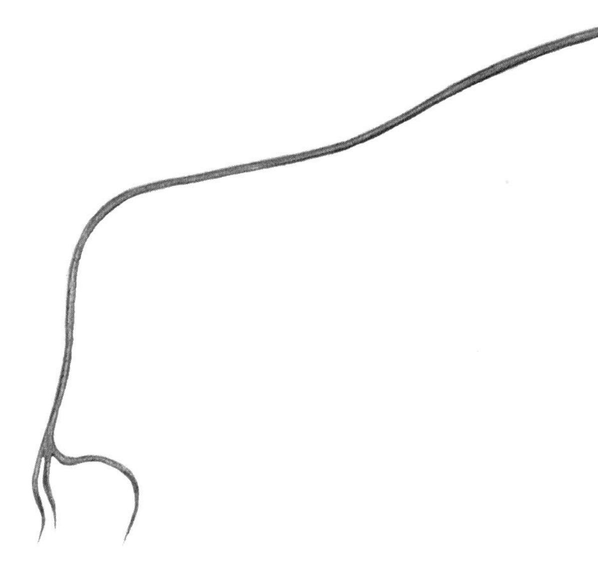

Smoked Chicken with Three Gourds

If you can find a cold-smoked chicken, ready for roasting, that is the ideal. Roast it, basting with a glass of dry sherry, then remove and keep warm. Pour fat from the pan, add 250 ml (8 oz) giblet or poultry stock and boil down to make a little sauce. Thicken it if you like by swirling in a few cubes of butter – or leave it alone.

If all you can get is hot-smoked chicken, ready for eating, give it half an hour in a self-basting roaster or chicken brick in a moderate to hot oven.

Meanwhile prepare and slice across 4 small to medium bitter gourds, halve longways 10–12 very small courgettes, and cut half a cucumber down into very thin long slices. Finely chop a clove of garlic, and, separately, about 3 tbsps mixed parsley, chives and coriander or tarragon leaves.

Blanch the bitter gourd for 3 minutes in boiling salted water, then finish in a little butter with some of the garlic. Cook the courgettes in a little butter in a covered pan. Fry the cucumber briefly in butter. Arrange the 3 gourds to one side of a hot serving dish. Cut up and arrange the chicken beside them. Scatter with the remaining garlic and herbs mixed together. Serve sauce separately and provide bread or basmati rice.

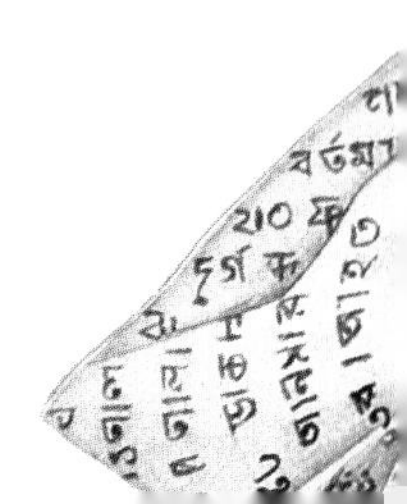

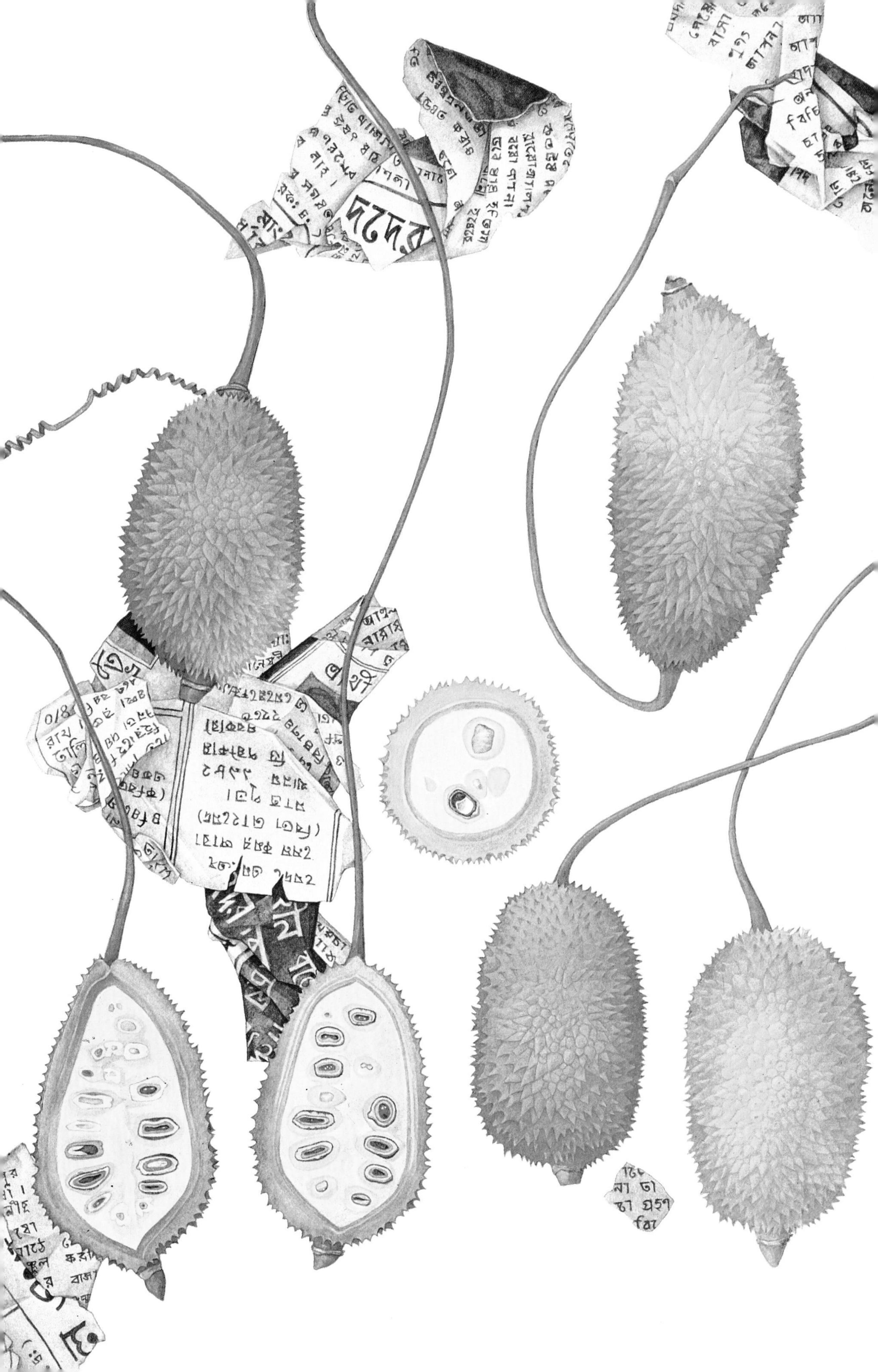

BOTTLE GOURD

(Lagenaria siceraria)

Remains of the bottle gourd have been found both in Mexican caves dating from 7000 BC and in Egyptian tombs of the fourth millennium BC. No other crop was common to both the Old and New Worlds before the age of Columbus, so researchers have hypothesized that the seeds must have drifted by sea from Africa to the Brazilian coast.

You will probably find bottle gourds called dudi or lokhi in Indian shops.

CHAYOTE

(Sechium edule)

The chayote, chocho, choko, christophine or – in Louisiana – the mirliton, à cheerful twiddly word that belongs more to the refrain of some old French song than to a gourd, is also more flatly known as the vegetable pear. This describes the kind of thing to look for.

Not that I would suggest looking for it very hard or long or persistently. It is pleasant enough, somewhere between a cucumber and marrow in texture, without a very distinct flavour, a vehicle rather for other flavours and textures than a delight in itself. Though I must admit it is popular in many of the hot parts of the world where it now grows, having spread from Mexico, where the Aztecs cultivated it and called it *chayotl*.

The secret of its success must be its easy rampageous growth. If you leave one to sprout in a pot, it will take over your kitchen, your house, your life, until you are embowered like the princess in the fairy tale. Grown outside it makes the ideal terrace covering. Make a little scoop of a cradle in the earth at the foot of the iron or wood uprights, and lay a chayote in each one, narrow end slightly up. Then, with luck and a good summer, they will cover the whole framework decorously and the fruits will hang down, tiny, pale green, almost translucent pears, baubles for a summer festival.

The Chinese call chayote Buddha's hand gourd, as if between moments of meditation in a summer bower, he reached up an elegant, formal hand to pick some for a frugal meal. An Australian friend assures me that, like courgettes, it is at the miniature stage that chayote are most worth eating. Pack them into a generously buttered pan in a single layer, cover them and get them going over a gentle heat. Salt and serve them when they are tender-crisp.

To Choose and Prepare

It does not matter whether chayote are smooth or whiskered with little spines, whether they are whitish or a clear spring green, they are all cooked in the same way. They are usually pear-sized. Avoid any that are tired and wrinkled, bruised or damaged. They have undulations and runnels in them which can make peeling difficult. Luckily they do not require peeling until after they are cooked for most recipes, if at all.

Inside the chayote has a round, flat seed that is nutty to eat, slightly crisp (after cooking, that is). Most cooks regard it as the parson's nose, the best part, too small to be shared round, and so reserved without guilt as a *bonne bouche* for personal consumption. If honour does not allow this, it can of course be chopped and incorporated into the recipe.

For people who can grow chayote easily, it may be useful to point out that its young shoots are cooked and eaten like asparagus. The young leaves are also edible, and so is the two-year-old tuber.

Such details are not of much use to northern cooks, who will find it easy enough to adapt cucumber, courgette or marrow recipes for the fruit, which is the only part they are likely to come across. It can be cooked and finished in the summer manner, with butter, cream and a dust of nutmeg, as a partner for salmon and sole. Generally speaking, though, crispness, piquancy and spiciness characterize most of the preparations given for it in books of tropical and sub-tropical cookery, once they get beyond the basic universal simplicity of boiled or steamed chayote, buttered and served with coarsely ground pepper: to me this has all the attraction of school marrow.

A better way is to dip cooked slices into egg and breadcrumbs or into batter and fry them crisply – crispness counteracting the watery mildness of the flesh; or simmer them in a light, fresh tomato sauce, seasoned with a small hot chilli or cayenne pepper.

Cooked chayote can also be cubed and fried with other vegetables to make a Spanish tortilla or Persian kuku. One recipe given by Elisabeth Lambert Ortiz has cubed fried chayote moistened with a light tomato sauce including a hot chilli, and scrambled with beaten egg rather than set smooth as in tortilla or kuku.

Stuffed Louisiana Mirlitons

Halve cooked chayote. Scoop out flesh and seed, leaving a good 'wall'. Chop flesh and seed and add to a shrimp stuffing, or make moussaka-style stuffing, and pile into the chayote. Sprinkle with breadcrumbs and cheese and then melted butter. Bake at gas mark 5–6, 190–200°C (375–400°F) for about 35 minutes until lightly browned.

Chayote Salad

Put cooked, peeled, warm sections of chayote into an olive oil vinaigrette, lightly flavoured with Dijon mustard and chopped parsley, chives and chervil. When ready to serve, drain off excess moisture and scatter with tiny bread dice fried crisp in olive oil.

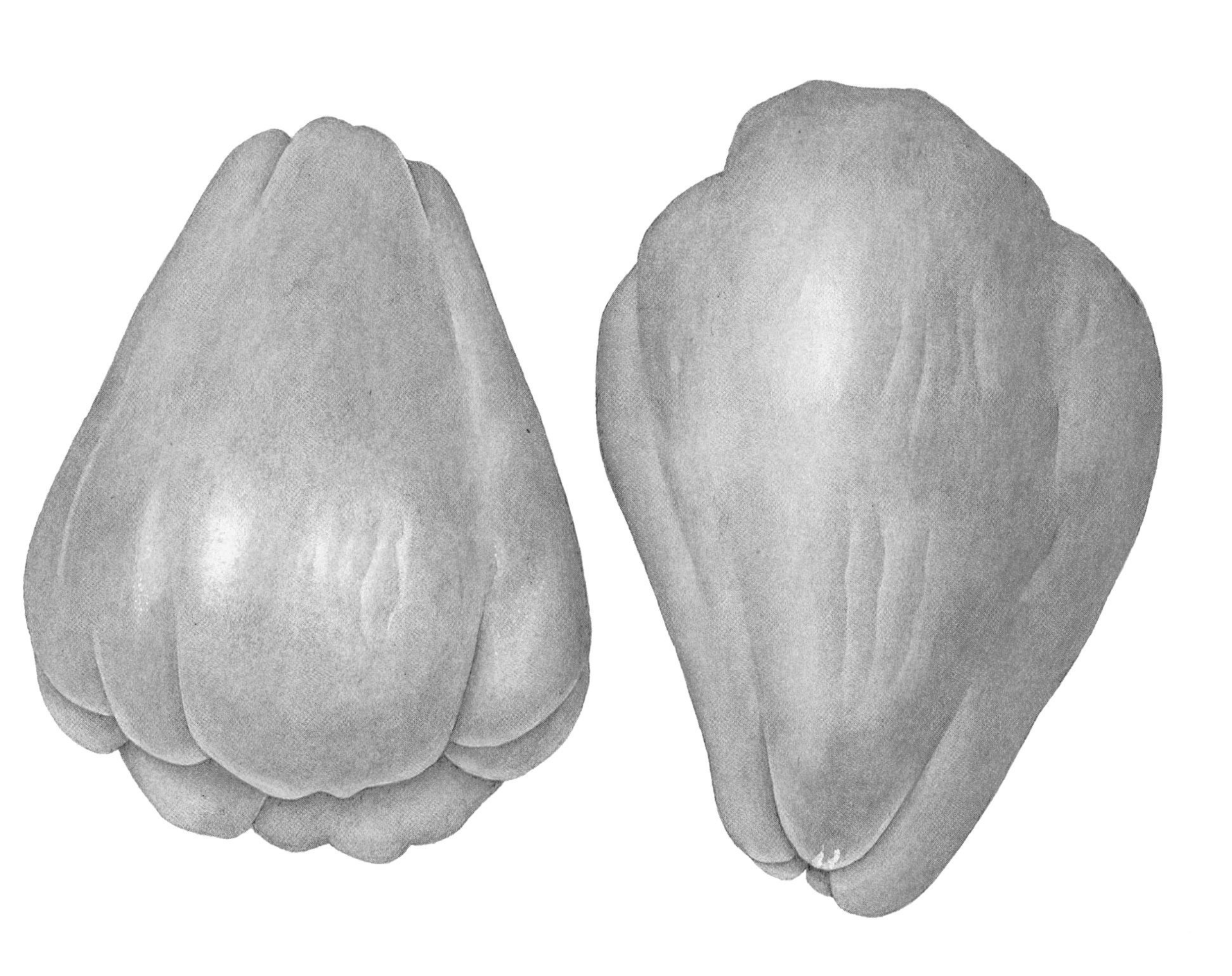

IVY GOURD or TINDORI or TINDOLA

(Coccinia grandis)

John Organ observes with justice that if you call a gourd a gourd, people think of it only as an ornamental or useful vegetable oddity, gourds to make dishes, ladles and bottles for distant tribes somewhere in the tropics that they cannot name (in fact gourd is a general term for generally hard-shelled *cucurbitaceae*). Call a gourd a squash and people realize that it is or may be edible – squash being more of a wooden-spoon and potato-masher word. Certainly the tiny ivy gourds that you see occasionally in our Indian and West Indian markets look far more ornamental than edible: when you cut them

in half before cooking them, the tight-packed seeds crammed into the thinnish skin make you wonder how on earth such things will go down. What is there actually to eat?

I would say that the point of these little gourds as food is precisely this crammed regularity of seeds, their crunchiness having quite a different quality from, say, the softer, nut-like, giant seed of the chayote. It comes in very welcome as a contrast to the watery flesh of other gourds.

Although the box of tumbled ivy gourd looks appealing, it gives one no idea of the exuberance of the plant that rampages over hedgerows and arbours and trellises. First come the white flowers, almost as large as the fruit, very noticeable like the open mouth of a trumpet and bursting out into five petal points; then the patterned green fruit, about 5 cm (2 in) long, with a bursting plump look, slightly shiny, which eventually ripens to a flaunting scarlet. We do not see the fruit at this brilliant stage, but in its homeland of the North African and Asian tropics it is eaten as a fruit. It attracts the birds, too, which peck enthusiastically at the colour and soon get through to the seeds – and so they are rapidly distributed and the plant continues its vigorous climbing without much need of man.

To Choose and Prepare

Choose tight-fitting gourds that are sheeny and smooth, not broken at all or bruised. At home, cut them in 2 downwards from the stalk end, sprinkle them with a little salt and leave them in a colander for half an hour, or longer. Rinse off and dry them, then use them alone or with other vegetables, with or without meat or fish, in any of the curried dishes given elsewhere. If you want to prepare them Western-style, sweat them gently in butter or olive oil in a covered pan, and add skinned, chopped tomato to make a little chunky sauce. Add salt and pepper, a pinch of sugar if the tomatoes are pallid, and some torn basil to finish with. In other words treat them as if they were halved courgettes. A good vegetable on its own, or with some bland meat that will contrast with the sharpness of the tomato and the crunch of the seeds.

Sabzi Dam or Steamed Vegetables

Although Jack Santa Maria does not make much use of the unusual tropical gourds and yams in his books on Indian food, many of his recipes can easily be adapted to them. Here is one which makes use of courgettes and either marrow or pumpkin: ivy gourd or loofahs and snake gourd do well instead. To a Westerner this technique of steaming vegetables with a yoghurt paste seems revolutionary.

1 tsp turmeric
2 tbsps mint leaves, roughly chopped
1½ cm (½ in) piece ginger, peeled, sliced
4 cloves garlic, skinned, halved
juice of 1 lemon
250 ml (8–9 fl oz) yoghurt
250 g (8 oz) ivy gourd, halved and salted, or courgettes or loofah, chopped
250 g (8 oz) marrow or pumpkin or snake gourd, peeled, seeded, cubed
250 g (8 oz) white cabbage or cauliflower, chopped
250 g (8 oz) mushrooms, sliced
1–2 tbsps ghee or melted butter
1 medium onion, finely chopped

Pound together the first five ingredients and mix in the yoghurt. Put the vegetables in a dish and cover with this paste. Leave them to marinate while you heat the ghee or butter and fry the onion lightly. Add 250 ml (8–9 fl oz) water, or vegetable stock if you have it, and bring to the boil. Tip in the vegetables with the yoghurt mixture, bring to simmering point, cover and leave until tender. Check from time to time and replenish the water or stock if the vegetables show a tendency to stick. Serve hot.

SNAKE GOURD

(*Trichosanthes cucumerina* var. anguina)

Of all the tribe of gourds, the snake or serpent gourd, also known as chichinda, is the most ingenious. It grows long – and at first green streaked with greenish white – undulating into curves and curls, taking bizarre twists. By the time it finishes, if it is left alone to get on with its destiny, it can measure 2 m (2 yds), by which time it has turned yellow to red and is quite uneatable. Inside the seeds are neatly packed into a red spongy pulp. This plant of the lowland tropics, of India in particular which is its original home, has another more elegant peculiarity – flowers fringed with fine hairs (*Tricosanthes* means hairy flower), very delicate and white. They open in the early evening and stay open all night, breathing sweetly just as Marvel of Peru flowers do, or tobacco flowers, in our northern gardens.

I imagine – for I have never seen the snake gourd growing – some terrace in the angle of a low house, a bamboo-trellis framework shaded over with the creeping vine of the snake gourd, and other gourds as well, the flowers scenting the air in the darkness, the young fruit hanging down and the owner going from one to another tying a little stone to the end of each one, a pendant to curb the ebullient growth. Some of the fruit may be left to amuse people with their odd shapes, most will be picked when they are still very green, and certainly no longer than 60 cm (2 ft). These will be cut for the pot, or for the market.

It must be admitted that the flavour of the snake gourd is not up to its splendid eccentricities of flower and fruit. No rival to melons, cucumber and pumpkin, nor to the supple bitterness of the bitter gourd. It makes a pleasant enough dish when mixed with other items, or well spiced in the Indian fashion. Loofah recipes can easily be adapted to the snake gourd – the Sulawesi manner of stuffing and baking them on p. 56 is good – so can Western recipes for courgettes and cooking cucumbers. If you try growing it yourself as Philip Miller did in the middle of the eighteenth century, I suppose in Chelsea Physick Garden which was his kingdom, you will remember the flower and the odd behaviour of the fruit much longer than the taste.

To Choose and Prepare

You may not have much choice in the buying of snake gourds, but make sure they are firm and bright. If you intend to stuff and bake them, go for the shorter, straight ones. Peel them thinly in any case, or scrape off the tough part, then slice them across into pieces and remove the seeds and pulp.

Snake Gourd Malay Style

This is a fairly substantial dish, and like much Malay food far hotter than Nepali or Punjabi vegetable dishes.

2 long snake gourds, peeled, seeded
1 tsp turmeric
250 g (8 oz) yellow dal *(red lentils)*
salt
ghee or sunflower oil
5 medium onions, sliced
¼ tsp cumin seeds
½ tsp mustard seeds
5 green chillies, seeded, sliced
2 stalks curry leaves

Slice the gourds fairly finely and sprinkle with the turmeric. Half-cook the lentils in water to cover generously, then add the gourd and complete the cooking. Season with salt.

Melt just enough ghee or oil to cover the base of a sauté pan and cook the onions in it gently to soften them. Add the remaining ingredients and continue to cook for 3 minutes, then stir in the *dal* and gourd and give

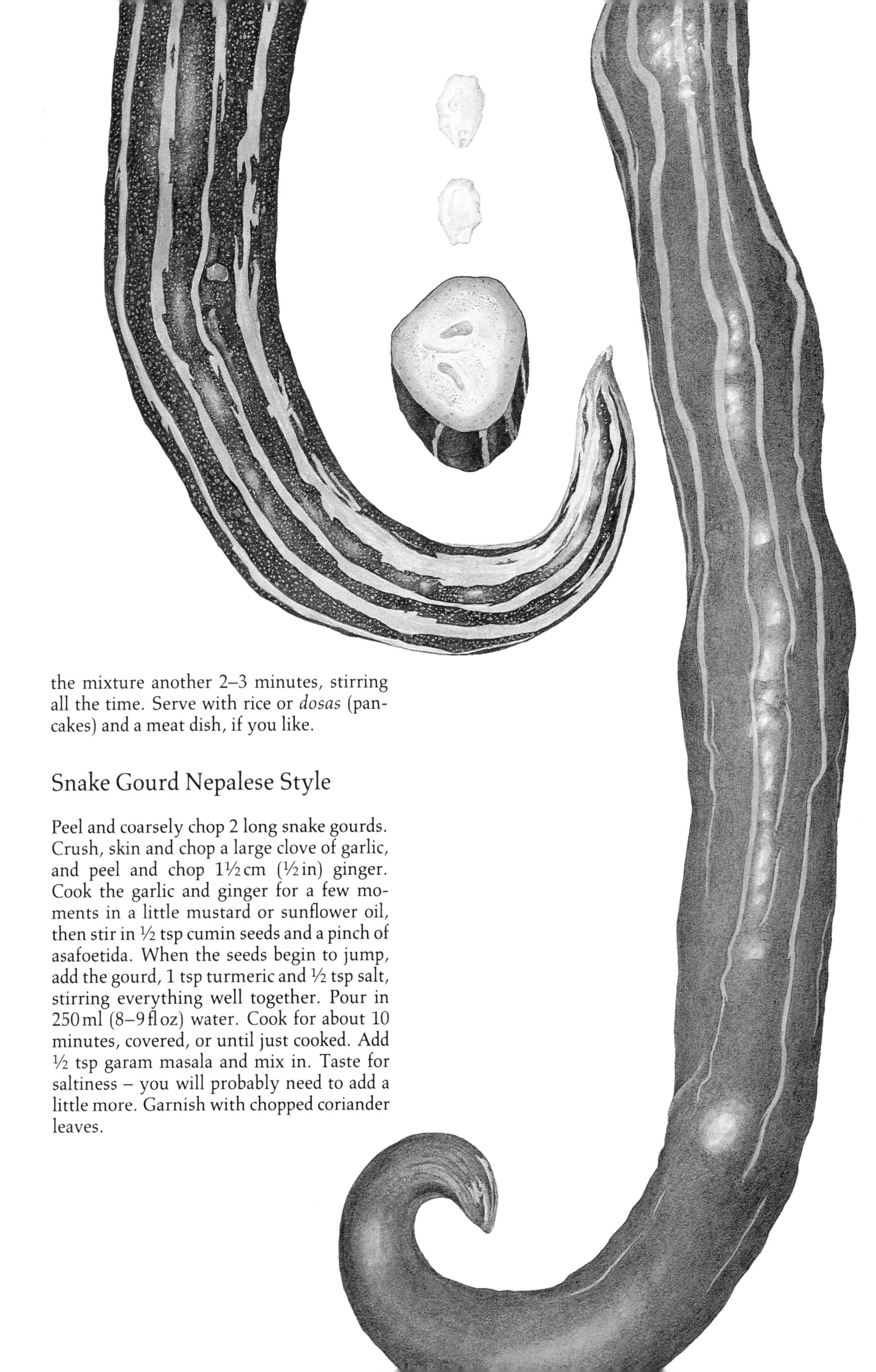

the mixture another 2–3 minutes, stirring all the time. Serve with rice or *dosas* (pancakes) and a meat dish, if you like.

Snake Gourd Nepalese Style

Peel and coarsely chop 2 long snake gourds. Crush, skin and chop a large clove of garlic, and peel and chop 1½cm (½in) ginger. Cook the garlic and ginger for a few moments in a little mustard or sunflower oil, then stir in ½ tsp cumin seeds and a pinch of asafoetida. When the seeds begin to jump, add the gourd, 1 tsp turmeric and ½ tsp salt, stirring everything well together. Pour in 250ml (8–9fl oz) water. Cook for about 10 minutes, covered, or until just cooked. Add ½ tsp garam masala and mix in. Taste for saltiness – you will probably need to add a little more. Garnish with chopped coriander leaves.

OKRA

(Hibiscus esculentus)

Lady's fingers is our English name for the soft tapering pods known as okra elsewhere. It reminds me of a description I once read by a French traveller of Creole ladies in the Antilles eating with their fingers. They picked up their food with such lazy but skilful grace that not a drop fell on to their pale dresses. Their white hands remained clean.

In fact the name is shortened from Our Lady's fingers, by analogy with Lady's mantle or Lady's bedstraw, an image of a grander, less indolent softness. The general idea, though, is worth remembering as a market guide, since the pods should be young and tender. Once the angle lines are brown and dark patches appear, they are not worth buying. They will cook to a kind of stringiness that winds round your teeth and sticks in the throat.

Practicality apart, I go for the name of okra because it reminds me of the plant's native land. And of its history, which is not soft or pretty at all. Okra comes from *nkurama,* its name in the Twi language of the Gold Coast. Slaves took okra with them to the Caribbean and the southern states of America. The name that sticks there is another African word, gumbo, of Angolan origin. And gumbo is the name given to American dishes in which it appears, even in small quantities, because the pods contain a clear gummy liquid which smooths the sauce. It thickens it too in a jellied way which makes you lick your lips. So you may get shrimp gumbo or chicken gumbo, stews and soups which have this characteristic texture, even though the sliced green pods have disappeared from view into the general *mélange.*

To Prepare

Nip off any stalks. Trim round the tiny cone at the stalk end to pare off the hardness. Be careful not to pierce this cone if you intend to keep the okra whole, seeds and juice inside. Should there be dark lines down the angle of the okra, slice them carefully away using a potato peeler for a shallow, even cut.

Bamies Meh Domates (Greece)

1 large onion, chopped small
3–4 tbsps olive oil
½ kg (1 lb) okra, prepared
1 large clove garlic, chopped
½ kg (1 lb) tomatoes, skinned, chopped
juice of 1 lemon
salt, pepper, sugar, rigani (or origano)
chopped parsley

Soften the onion in the oil without colouring it. Push it to one side, raise the heat slightly and brown the okra lightly. Mix the 2 and spread in a single layer in the pan. Put garlic and tomato on top with lemon juice and seasonings. Add a little water to prevent catching. Cook gently for about 20 minutes until the okra is tender and the tomatoes reduced to a sauce. Do not stir: shake the pan instead. Check seasoning. Serve hot, cold or warm, sprinkled with plenty of parsley, as a first course or vegetable.

African Okra and Shrimp Soup

250 g (8 oz) prawns or shrimp, shelled
250 g (8 oz) okra, prepared
300 ml (10 fl oz) water
1 small tomato, skinned, chopped
½ level tsp hot chilli powder
2 tbsps best-quality palm oil
salt
1 ground seed African nutmeg

Put prawns or shrimp into a processor, with half the okra, and slice the rest into a pan. Process the shellfish and okra to a paste (or use a blender or pestle and mortar), add to the pan, and then add the water, tomato, chilli powder, palm oil, salt and nutmeg (or a grating of our familiar nutmeg). Simmer for 6–7 minutes, by which time the soup should be ready – check that the sliced okra is tender. Season to taste. Serve with slices of boiled green plantain, or with boiled yam. Groundnut oil could be substituted for the palm oil (see p. 99), but the flavour will not be quite the same.

Chicken and Oyster Gumbo

1 chicken, cut into 8 pieces
250 g (8 oz) cubed gammon
1 medium to large onion, chopped
1 chopped sweet red pepper or seeded hot red chilli
375 g (12 oz) sliced okra
chicken fat and/or lard
2 large tomatoes, skinned, chopped
stock to cover
bouquet garni
2 cloves garlic, chopped
cayenne pepper or Tabasco sauce (see recipe)
12–24 oysters or mussels
chopped parsley

Brown first five ingredients lightly in the fat. Pour off surplus. Add tomatoes and stock to cover. Tuck in *bouquet*, scatter on garlic and add cayenne or Tabasco if you used the sweet red pepper. As the chicken simmers, open oysters or mussels in the usual way, keeping their liquor and straining it. Add the liquor to the chicken, gradually, to taste. When the meat is tender, check the seasoning and remove the *bouquet*. Add oysters or mussels. Keep heat below boiling point for 5 minutes, add parsley and serve with rice.

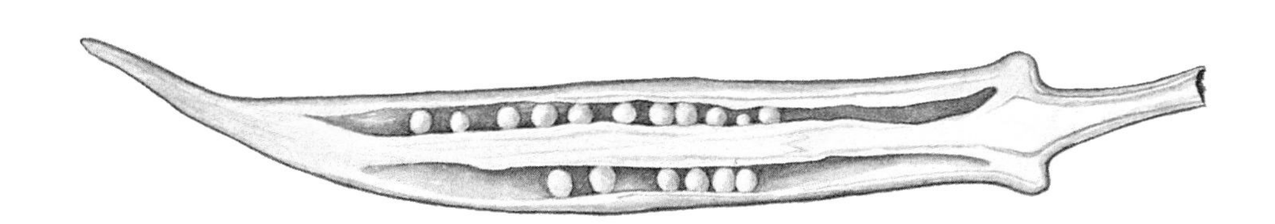

PLANTAIN *(Musa paradisaica)* and BANANA FLOWER or HEART *(Musa sp.)*

Plantains are a kind of cooking banana. They have a more masculine look than our familiar dessert bananas, being firmer and tighter to their skins, less yielding as they ripen. You can eat them raw when they turn yellow, then black, but they are stodgy this way, not succulent or honeyed like a truly ripe banana that melts in your mouth. They need to be cooked to be enjoyable. On the whole they are eaten as a vegetable, providing a starchy background, a filler, to accompany stews of more expensive items such as meat and fish. Green bananas should also be cooked (unless of course you intend to keep them for eventual eating, when they turn golden yellow): use any of the plantain recipes below, or in other cookery books, cutting down the cooking time.

As you read through this book you will see that we are limited in Northern Europe and America in the matter of starchy vegetables. We have taken up potatoes, though this was a slow business at first even with the Irish, who are thought of as the great potato eaters. Parsnips and Jerusalem artichokes, salsify and scorzonera come far behind potatoes in our diet, no doubt because they are too positive in flavour. We have been rich enough for centuries to afford meat and fish as main items, with a little bread and nowadays potatoes. We have not looked much for starchy fillers. Presumably this is why so few of the more solid vegetables of the tropics and sub-tropics are known to us, whereas we have long been familiar with much of the fruit.

To Prepare

Plantains are not as easy to skin as you would expect. For slices, cut the plantain across at appropriate intervals and soak the pieces in salted water for about 30 minutes, or until you can push the edible discs from their close rings of skin. If you need the length of the plantain, or halves or thirds, cut down each angle, not too deeply, and pull away the strips of skin between, using a small knife to ease them away at first. The riper the plantain, the easier it is. Keep the pieces in salted or acidulated water until required, or they will discolour.

To Cook

Plantain pieces can be boiled like potatoes and need about 30 minutes. Inevitably their sweet starchiness is popular with spicy and chilli-hot stews of meat and beans, or with curries. Plantains can be grated and cooked in stock as a thickener of soups. Simplest of all, they can be baked or grilled in their skins: when they are soft to the touch, they are ready to eat.

Plantain Crisps and Chips

For crisps, choose yellowish green to yellow plantains. Halve them across the centre, peel and cut them down into long finger-like strips or ribbons. Dry them on kitchen paper and bend them into loops or curves (Nigerians are clever at this). Deep-fry at about 170–180°C (340°F) until golden. Sprinkle with salt and serve with drinks, or with white fish and poultry or bland meats.

Chips from Latin America are made by cutting the plantains into pieces 3–4 cm (1¼–1½ in) long. Deep-fry at a slightly lower temperature until they are cooked but not crisp. Drain on kitchen paper, then punch down with your fist until they are half as thick. Alternatively, overlap one thick slice with another and punch them both down – Elisabeth Lambert Ortiz recalls that she's seen cooks in Venezuela, on the coast, using large stones from the beach to do the flattening for them. Then deep-fry until the edges are crisp, the thick centres of the figure-of-eight shapes tender.

Plantain and Honey Pudding

Choose ripe plantains. Cut them across diagonally into thin slices. Deep-fry until crisp. Cool and serve mixed with toasted, slivered almonds and honey-sweetened whipped cream.

Chicken in the Kitchen Garden (Pollos en Huerto)

Elisabeth Ortiz who gives this old recipe in her *Complete Book of Mexican Cooking*, says that it reminds her of the ornate altars at Tepotzotlan – 'so overdone in decoration that somehow it works'.

1 2kg (4lb) chicken, jointed, seasoned
4 tbsps lard
2 onions, sliced
2 cloves garlic
500g (1lb) tomatoes, skinned, seeded, chopped
bay leaf tied with 3 sprigs of parsley
12 annatto seeds, ground
1/8 level tsp each ground cloves and cinnamon
125g (4oz) each peeled, sliced carrots, sweet potatoes and potatoes
1/2 l (18fl oz) dry white wine
125g (4oz) each sliced courgettes, shelled peas and cut green beans
1 ripe plantain, peeled and sliced, or 1 large firm banana
2 cooking apples, cored, peeled, cut up
2 peaches, stoned, peeled, sliced
2 pears, cored, peeled, sliced
2 quinces, peeled, sliced, cored (optional)
4 slices pineapple, peeled, cut in chunks
6 large ripe olives
2 level tbsps drained capers
30g (1oz) raisins
3 or more jalapeno chillies, seeded, rinsed and cut in strips

Brown the chicken lightly in the lard. Put into a huge flameproof pot. Cook the onions gently in the remaining fat, and as they turn transparent add the garlic. When the onions are completely transparent, scoop them and garlic into the pot. Add ingredients down to and including wine. Cover and simmer gently for half an hour. Remove the breast if it is done. Put in remaining ingredients and cook over a low heat for a further half hour, or until the chicken is tender. If necessary, return the breast pieces to heat through.

Serves 8.

BANANA FLOWER or HEART

This is the male part of the flower, a rounded spike. To get at the edible part, you cut away the outer reddish petals until you get to a paler inside, which is firm but tender enough to eat.

In Cambodia they cut this up into thin slices and make a salad. Other ingredients are pounded dried shrimp, chopped crisply cooked bacon skin, some tbsps of chopped cooked pork. The dressing is thick coconut milk, brought to the boil, then cooled and seasoned with *nuoc mam* (fish sauce), lemon juice, a little sugar and a paste made by browning together garlic, fennel root, allspice and shallot *(kapik phat)*.

Sri Owen gives a simpler recipe for an everyday sort of dish in Alor (Nusa Tenggara Timur), 'where of course you can pick your jantung pisang' i.e. banana flower 'whenever you like in your back garden'. You boil the pale inner part whole for 15 minutes, then cool it. Put 250ml (8fl oz) thick coconut milk (see p. 39) in a pan with 3 sliced shallots and some salt and pepper. Bring them very slowly to the boil while you slice the banana heart: cut it into four lengthways, then slice each piece across into thin flat quarter-rounds. Add to the coconut milk when it boils and simmer 10 minutes. Stir often so that the milk does not curdle and keep the heat below boiling point. Serve hot with meat or fish and of course with rice.

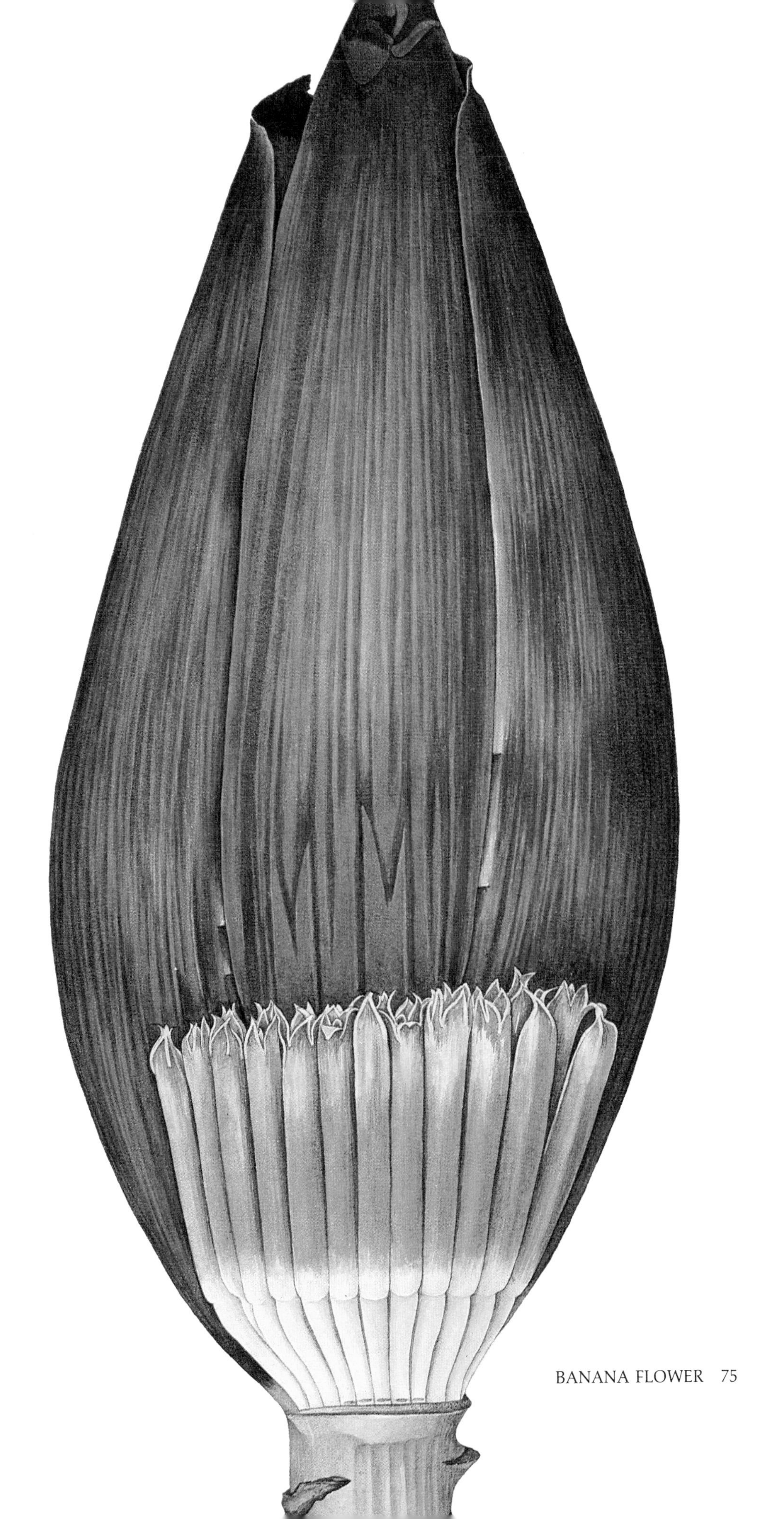

BANANA FLOWER

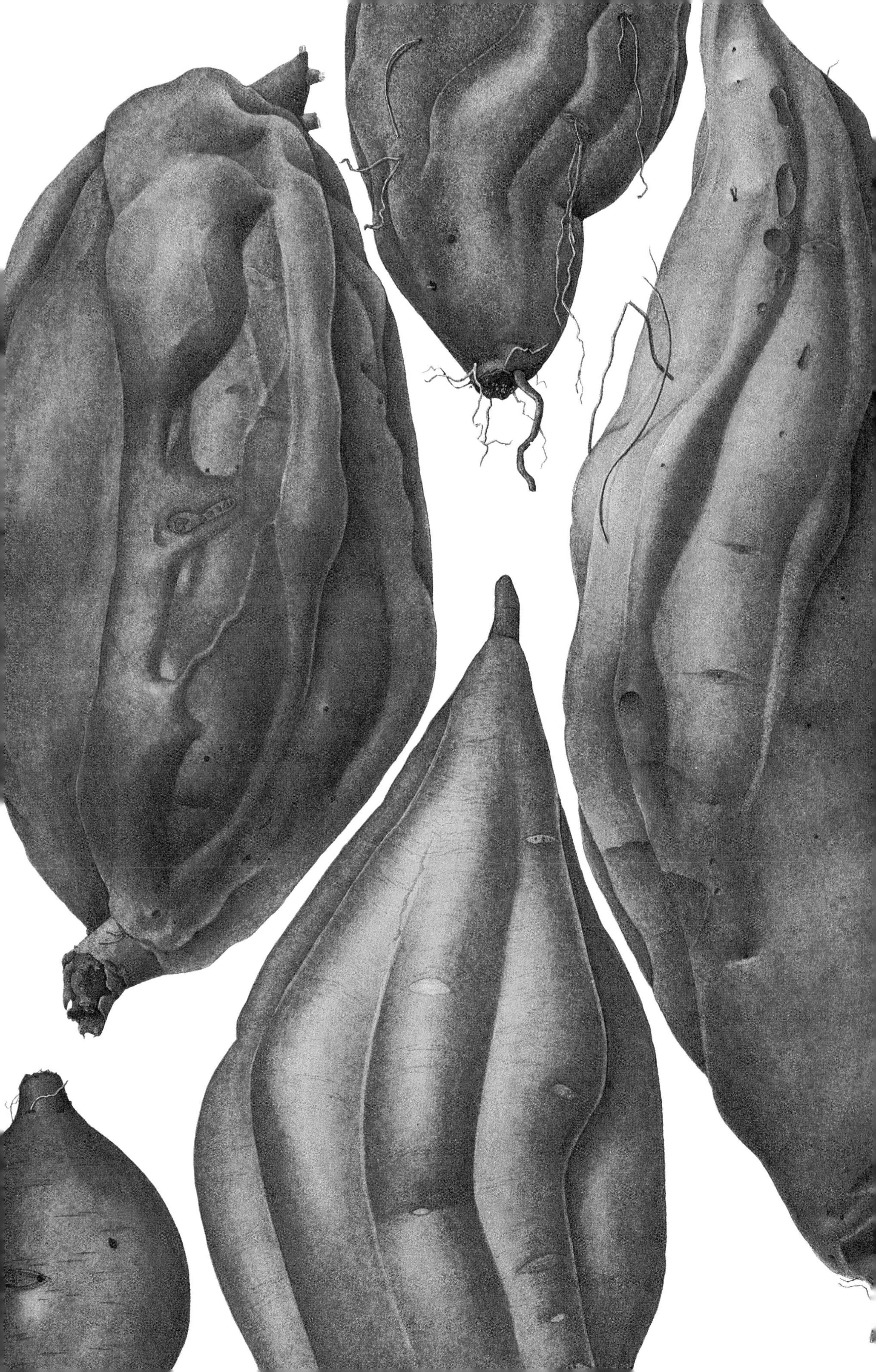

SWEET POTATO

(Ipomoea batatas)

Christopher Columbus first encountered the sweet potato in Haiti in 1492, and brought it back to Spain the following year in the collection of booty from his New World across the Atlantic. Sweet potatoes rapidly became an important part of the provisioning of ships coming from Haiti – they are now the major root of the Old World tropics – and soon they were growing in the warm climate of Spain. They were in fact the potatoes of Europe for the next sixty years, the name coming from the Arawak *batatas*. Our present-day potato, *Solanum tuberosum* from the colder Andes, reached Spain sometime in the 1560s, but did not get to Britain until the 1580s. This newcomer could be grown in our northern mists very successfully, and gradually it usurped the name, *Ipomoea batatas* becoming the Spanish potato that would always be an import and therefore an expensive luxury by comparison. The first record of the name sweet potato is in the *Oxford English Dictionary* of 1775.

All this may seem pedantic, but anyone who likes to try out dishes from seventeenth- and eighteenth-century recipe books should bear it in mind and ponder whether those wonderful pies of many ingredients including potatoes, or those potato cakes with sugar, ought not in fact to be made with sweet potatoes rather than Desirée, Maris Piper or King Edwards.

What puzzles me is why our potato, our common potato, should have been given the same name in the first place. It is not like the sweet potato at all and tastes quite different. Even the whiter sweet potatoes are more like chestnuts to eat, though they have the look of our ordinary potato. In Genesis it tells how God made animals out of dust and brought them to Adam to see what he would call them – 'and whatsoever Adam called every living creature, that was his name'. Plants are not mentioned, neither are fish, and whoever got to work on them over the centuries in various parts of the world managed to make some intricate muddles.

Writing about *Ipomoea batatas* reminded me of the glorious blue-flowering fences and walls in our part of France, covered with the bright *ipomées* or morning glories that open early and are closed and over by midday. Whoever would think that they are related to sweet potatoes? But then I have never been daring enough to dig up anyone's precious roots of morning glory to check on this, or lucky enough to have seen sweet potatoes in flower.

To Choose and Prepare

There are many varieties of sweet potato with flesh varying from white through to deep orange. I prefer the white-fleshed kind which keeps firmer, more like chestnut, although I read that the orange-fleshed kind is more nutritious. Its softer, more melting flesh seems less attractive to me, but this is a matter of personal taste. To see what colour flesh sweet potatoes have – their skins are much the same pinkish-brown colour – scrape the skin off a little with your fingernail before you buy them.

To prepare them, give them a good scrub and cut off any dry, unpromising bits at each end. They can now be baked like the ordinary potato. Push a long skewer through to conduct heat to the inside, rub them with salt or wrap them in a butter paper, and put them into the oven preheated to gas mark 6, 200°C (400°F). The time required will depend on their thickness through, rather than their total weight.

Sweet potatoes can also be grilled. Cut them across into slices about 1 cm (good ¼ in) thick, brush them with melted butter or oil and put them under a preheated grill.

From their sweet similarity to chestnuts, sweet potatoes are often partnered with orange and much sugar to produce a candied

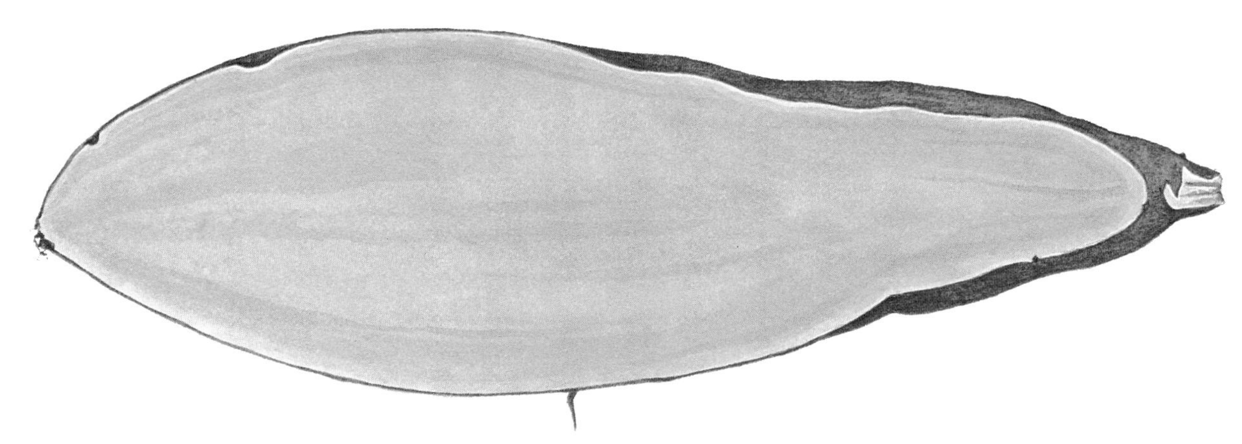

result that could be eaten as a dessert, but is more often served with ham. I find this rather sickening, even when the quantity of sugar demanded by American recipes is much reduced.

Sweet potato pulp, especially of the orange varieties, can be used instead of pumpkin to make open tarts.

Very thin slices can be deep-fried to make sweet potato crisps. Pan-fried sweet potatoes are good, too. Cook them gently in butter, covered, and turn them over after 5 minutes. Give them another 4 minutes and then remove the lid so that the watery juices can boil steadily to a glaze.

Carbonada Criolla

A magnificent dish from Argentina, ideal for a harvest celebration or Hallowe'en party. The correct gourd is really the Argentine *zapallo* (West Indian *calabaza*). Hubbard squash is the next best thing. I use an orange pumpkin of Cinderella's coach shape, and it works pretty well as long as you keep an eye on it in the late stages of cooking to see it does not burst (our pumpkin is rather more watery than the other squashes).

1 5 kg (10–11 lb) zapallo*, Hubbard squash or pumpkin*
125 g (4 oz) butter, melted
60 g (2 oz) sugar

Stew
1 kg (2 lbs) boned veal or chuck steak, cut in 2½ cm (1 in) cubes
oil
1 large onion, chopped
3 large cloves garlic, chopped
1 green pepper, seeded and coarsely chopped
2 hot red chillies, seeded and chopped
500 g (1 lb) tomatoes, skinned, seeded, chopped
tomato concentrate
¾ l (1¼ pts) veal, chicken or beef stock
¼ l (9 fl oz) dry white or red wine
bouquet garni
1 level tsp origano
salt, pepper
500 g (1 lb) white-fleshed sweet potatoes, peeled, cubed
500 g (1 lb) potatoes, peeled, cubed
500 g (1 lb) slice pumpkin, peeled, cubed
3 corn cobs, cut in 2½ cm (1 in) slices across
6 peaches, preferably white, skinned, stoned, halved
125 g (4 oz) long grain rice (optional)

Cut a lid from the gourd. Remove the cottony fibre and seeds. Brush butter over the inside, then sprinkle it with the sugar. Replace the lid and stand in a well-oiled pan or dish that can give support and will look all right for the table. Bake the gourd for 45 minutes in the oven preheated to gas mark 5, 190°C (375°F). Watch it attentively after 30 minutes. Test the flesh, which should still be a little firm.

For the stew, brown the meat in a little oil and put into a large pan. Then brown the onion lightly, cooking it slowly to melt it before it colours. Add the garlic, pepper and chillies. Stir for 1½ minutes, then add to the meat. Bring the tomatoes, a small tsp concentrate, the stock and wine to boiling point and add to the meat. Put in the *bouquet* and origano. Cover and stew for about 1 hour until the meat is almost tender. Now taste and season. Put in the sweet potato and potato cubes with the pumpkin, which will dissolve into the sauce and thicken it. Add more stock if necessary, or a little water. Simmer for 10 minutes, without a lid on the pan. Add the corn, peaches and rice, if used, and cook for a further 8 minutes. Check seasoning. Aim to end up with a thick stew.

Fill the pumpkin shell with the stew (leaving some in the pan for second helpings) and return it to the oven for 15 minutes. Keep an eye on things to avoid the pumpkin bursting. Ladle the stew out of the pumpkin into soup plates or bowls, scooping a little of the pumpkin flesh from the shell, but not enough to endanger its firmness.

Serves 8–10.

Louisiana Yam Pone

This tea-bread can also be made with the true sweet yam, or with Cyprus eddoe. It is a much more sympathetic recipe than some of the candied dishes, but then I have reduced the quantity of sugar by half.

Peel then grate 300g (10–11oz) orange-fleshed sweet potatoes coarsely into a pan. Add 110g (3½oz) vanilla sugar and cook for about 3 minutes, stirring. Cool and add ½ tsp true vanilla essence (not flavouring) if used.

Cream 140g (4½oz) sugar with 125ml (4floz) groundnut, sunflower or corn oil. Add 2 egg yolks, ¼ tsp powdered cinnamon, 1 tsp grated nutmeg, ½ tsp salt, and then 175g (6oz) flour gradually. Mix 1 tsp baking powder with 5 tbsps water and stir into the flour mixture with 125–175g (4–6oz) chopped nuts (excluding groundnuts), 45–60g (1½–2oz) raisins and 30g (1oz) desiccated coconut afterwards. Add to the sweet potato mixture. Fold in 2 stiffly beaten egg whites.

Have ready a loaf tin 23cm (9in) long, lined with baking parchment. Pour in the mixture. Bake in the oven preheated to gas mark 4, 180°C (350°F), for about an hour, or until cooked. Cool in the tin for 5 minutes, then remove and put on a cake rack. Peel off the paper when you can do so without hurting the cake.

Sweet Potato Cakes (Getuk Lindri)

These are the Indonesian equivalent of our Mont Blanc cakes, made with sweet potatoes, which do have a chestnut-like consistency and flavour when they are sweetened. You must be sure to buy white-fleshed sweet potatoes.

Wash and quarter 1kg (2lbs) sweet potatoes and boil them until they are tender. Drain and skin them, then mash until smooth. Put 60g (2oz) grated *gula Jawa* (which is sugar from the sweet juice of the coconut palm flower) or brown sugar, into a pan with 6 tbsps thick coconut milk (see p. 39). Heat until the gula or sugar is dissolved. Let it cool slightly, then mix into the sweet potato. Push it through a ricer so that it comes through in long separate strands on to a baking sheet lined with greaseproof or baking parchment. Shape them into small round buns or little oblongs about 7 × 5cm (3 × 2ins), then arrange them on a plate.

Grate 60g (2oz) fresh coconut, and put it into a bowl with a pinch of salt. Sprinkle this on top of the cakes before eating them.

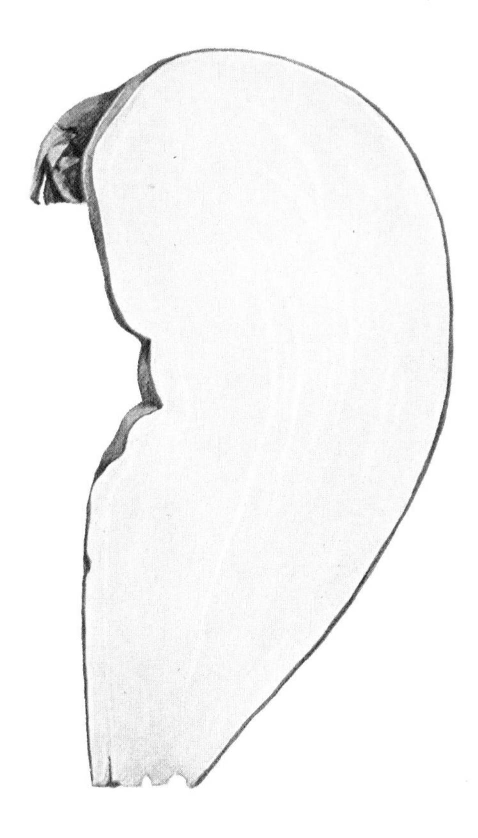

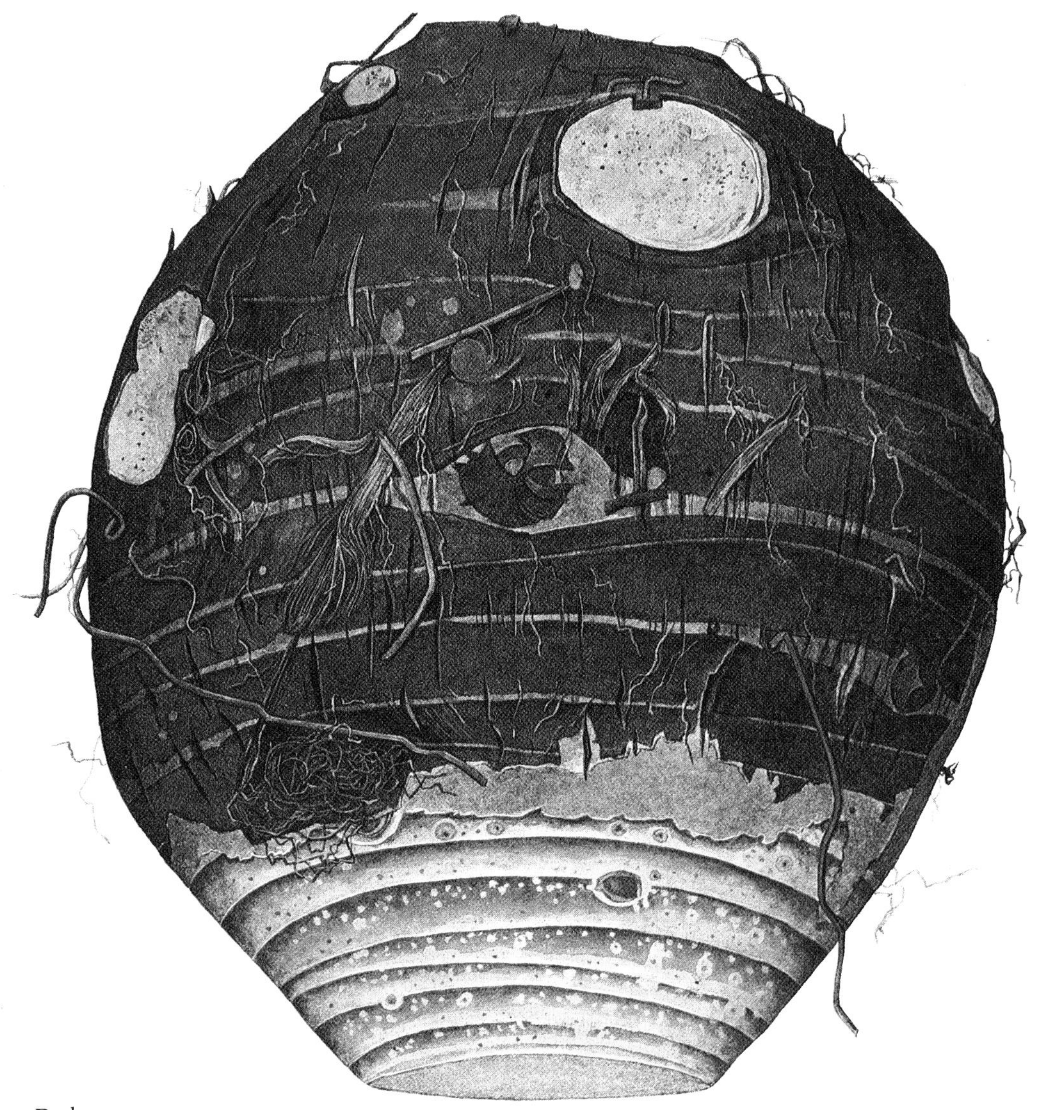

Dasheen

Eddoes

TARO

(Colocasia esculenta)

Taro is one of the many tropical cousins of our own odd lords and ladies. Unlike lords and ladies, which are decidedly inedible, taro is an important food plant in the hot regions of the world, in some parts a staple food, providing no less than four foods in one: huge leaves, which give the plant one of its names, elephant ears; the leaf-stalks and runners; the tender young shoots with barely unfurling leaves; and the starchy tubers, often called eddoes.

To its beauty and usefulness, you must add antiquity, since like the onion, taro is a plant man has eaten from so far back in time that it is unknown in a wild form. Most likely it originated in northern India, but there is no small, rough, tough version growing there today to show its primitive state or the size of the tubers that men first looked for and then took into patches of cleared ground to improve into the 200 and more varieties and cultivars of taro that are now known. Certainly it was taken to China early on and was well established by the Han dynasty: a stew made of venison and taro is mentioned on one of the bamboo slips found in the tomb of the Lady of Tai at Ma-wang-tui, whose body and silks were so remarkably preserved. With her were buried many foods and 312 inscribed bamboo slips listing things that were not represented among the grave goods and even giving information on the cooking of *keng*, which were soupy stews of meat or fish and vegetables, the main dish of Chinese eating from ancient times. Interestingly one of the many names that taro has acquired is *dasheen*, which is thought to come from 'da China' or 'de Chine'.

A good modern Chinese combination is duck with taro, not a *keng*, but a rich affair of deep-frying and steaming, and a stuffing aromatic with red bean paste, star anise, ginger and coriander all moderated and set off by slices of taro. Gloria Bley Miller gives a recipe in her *Thousand Recipe Chinese Cookbook* (also a source of ideas for other less common vegetables). The Japanese put taro into a mixed hotchpotch like *yosenabe*, or assemblies of vegetables. In the autumn at Kyoto there is a strange festival, the *Zuiki* festival, *zuiki* being their name for the leaf-stalks and runners of taro. The Temple god is carried about in grand state in a palanquin, gloriously decorated with fruit and vegetables, like a harvest festival. The roof that shelters him is thatched over with long green *zuiki* strands. Apparently when they are partly cooked and dried *zuiki* keep their goodness for a long time. One General in the sixteenth century, Kato Kiyomasa, had the *tatami* on the windy floor of his castle stuffed with *zuiki* rather than rice straw, to provide a last resort in a long siege.

Taro tubers have been sliced and dried or smoked for centuries as travelling food in south-west Asia. In cookery books today we read of delightful little Indonesian taro dishes, the tubers sliced thin and fried or boiled then finished in coconut milk or sugar, or luxurious snacks like *buntil*, in which a spicy paste bound with coconut milk is layered, wrapped and cooked in taro leaves, something like the Indian *patra* (see p. 88) though more complex in seasonings, but ancient reality was closer to that Japanese General's mats. Dried taro was an essential for survival on the long journeys that the Polynesians made from south-east Asia to the Pacific islands. These great sea migrations took place at the beginning of the Christian era. Seamanship apart, the extraordinary thing was that in those long open boats they managed to keep alive slips and cuttings of over 100 plants, including taro, bread-fruit, yams and the coconut palm.

One interesting kind of taro is the Cypriot colocassi. It grows fine and large, with a characteristic, easily identifiable stump. There it is teamed with pork or chicken, in a mixture not entirely different in the simplicity of its items from that Han stew of deer meat and taro, the *keng*, once served to the Lady of Tai. Of course they did not have tomatoes to sharpen and sweeten the mixture, since that particular encounter did not happen until after the discovery of the New World in 1492. Then taro crossed the Atlantic to the Caribbean and to Latin America and was made far more palatable with pepper sauces. Related plants were already being cultivated there, mainly species of *tannia* (this, from the Carib *taia*, is the name in Jamaica and other English-settled islands – Spanish names are *yautia* and *malanga*). Their leaves are often used for callaloo, that great Caribbean speciality associated in particular with Trinidad, though some insist that versions from the French islands are better.

What puzzles me about the world-round popularity of this food wherever it can be grown – and what makes me admire the hardihood of people who sorted out the edible plants from the almost edible, inedible and downright poisonous ones – is the daunting presence in taro of tiny crystals of calcium oxalate. They are an extremely unpleasant snag. My first experience with taro was cooking some of the leaf-stalks and runners. We boiled it with a little lemon juice, in short lengths, as instructed. It took on a rhubarb glow (sometimes it is compared in flavour with rhubarb). My daughter drained it and nibbled a bit. I nibbled a bit. All right but rather overcooked, we decided. Then our throats began to ache in a strange way, as if they were swelling up. We stood and looked at one another, too frightened to speak, fearing that we were soon going to be unable to breathe. After a lot of water the sensation vanished. A week or two later I tried some shoots of taro. The same thing happened. Even as I write about it, I can feel it still. What the original sensation of the first plant hunters must have been I cannot imagine, since our taro has been much tamed by selection (some kinds are quite without this unpleasantness). Apparently cattle will not touch the leaves at all. With the tubers you are safer since such crystals as they contain are peeled away with the skin.

I resent the lack of frankness about this in many of the cookery books that I possess. Even botanists do not quite come clean, sternly observing that you should be quite all right if the vegetable is 'properly cooked'. By which they mean boiling for an hour. Now this may be all right for tubers, but it is plain ridiculous for greenery because young shoots and leaves are ruined by such treatment. Only one person has really written frankly about this aspect of the taro, and about the sliminess of the tubers when cooked (no wonder the potato has become the worldwide popular root vegetable). That is Katherine Bazore in her *Hawaiian and Pacific Foods*, published in New York in 1940 with many subsequent printings – my copy, from the eighth printing, is dated 1960 (a book to look out for in second-hand shops and the catalogues of specialists in cookery books). She arrived at the University of Hawaii to teach home economics, and soon encountered *poi*, a pasty substance that can be made with bread-fruit, plantains and so on, but especially with taro. She wrote:

> How this thick, sticky, pale gray paste with no distinctive flavor could have become so popular with Hawaiians that in earlier days men and women are said to have eaten ten to fifteen pounds a day, is still a mystery. And I have eaten much *poi* since my introduction to it, too. I have learned to manage 'one finger', 'two finger', and even 'three finger' *poi*. One finger *poi* is so thick that enough for a mouthful clings to the forefinger when it is expertly twirled in the mass. Two finger *poi* is thinner and three finger is thinnest of all. It takes three fingers held tightly together and dipped into the mass to capture enough to make the dip into the *poi* worthwhile. I tried *poi* after it had fermented for a few days and had developed an acid flavour. But even then it needed the accompaniment of salted salmon, tomatoes, and onions – *lomi* salmon – to make the *poi* palatable.

To Choose and Prepare

With the green shoots, leaf-stalks, runners

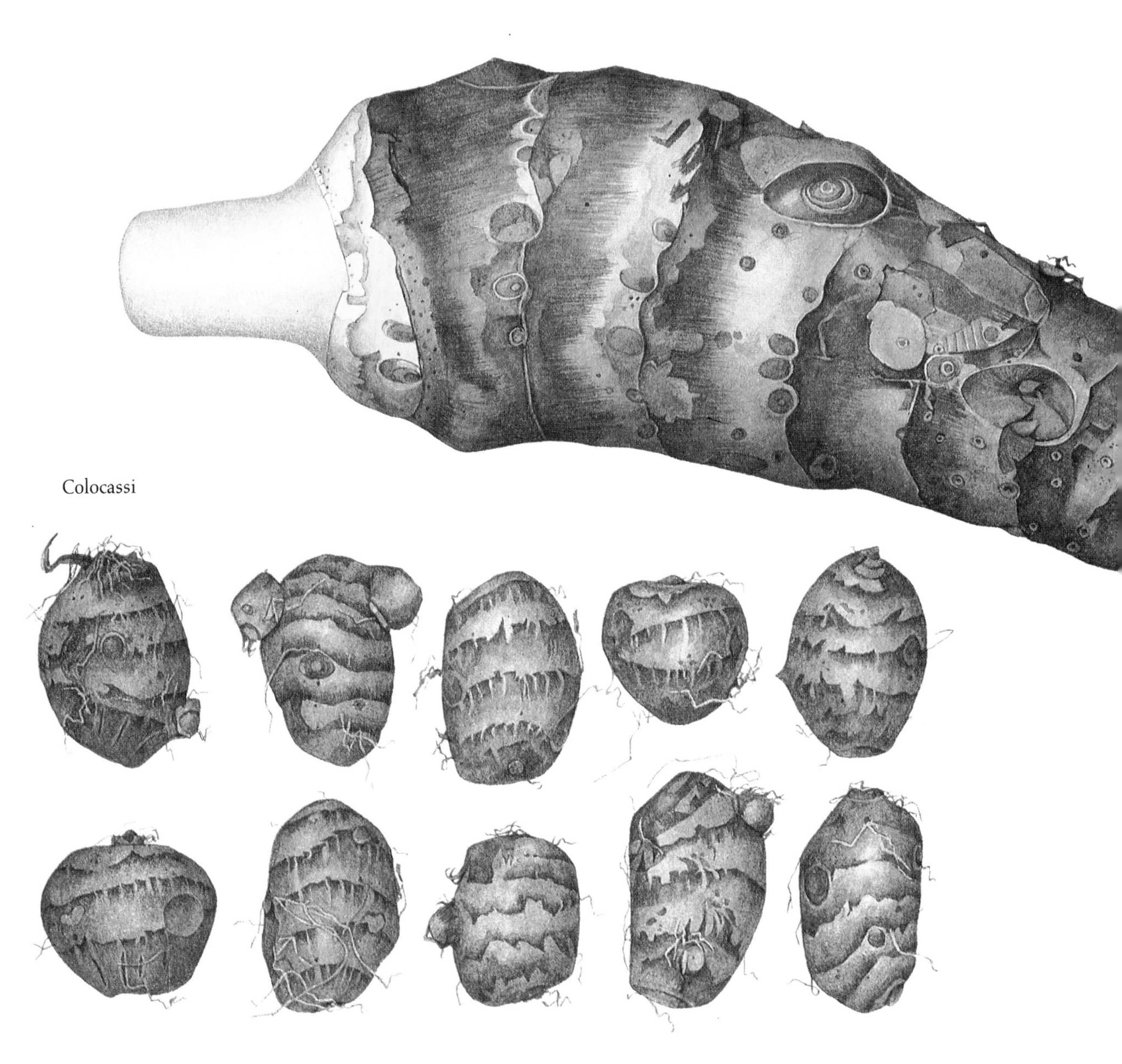

Colocassi

Eddoes

and leaves, apply the usual criteria for such things. Do they look lively or wilted? Nothing is worse than a truly dead vegetable. To prepare them for cooking, cast a knowing eye at the outer skin – does it need peeling away or does it look tender? Do the hard ribs under the leaves need removing partially? When you are washing the shoots, pay particular attention to the furled leaves. You may have to unwind them gently so that the water has a chance to lift away any lurking bits and pieces.

When it comes to the tubers, buy the size appropriate to the dish. Although they can be cooked like potatoes or yams, they are not so resilient and may collapse on you rather faster than you expect. If you must peel them, scrub all the dirt off, then dry them well and peel them without using any more water. Do not cut them up until they are required.

The green parts can be cooked like spinach, asparagus, French or yard-long beans, so long as you are able to buy varieties that do not sting the mouth. Cook a little bit and eat it yourself: wait 5 or 10 minutes to make sure there is no effect.

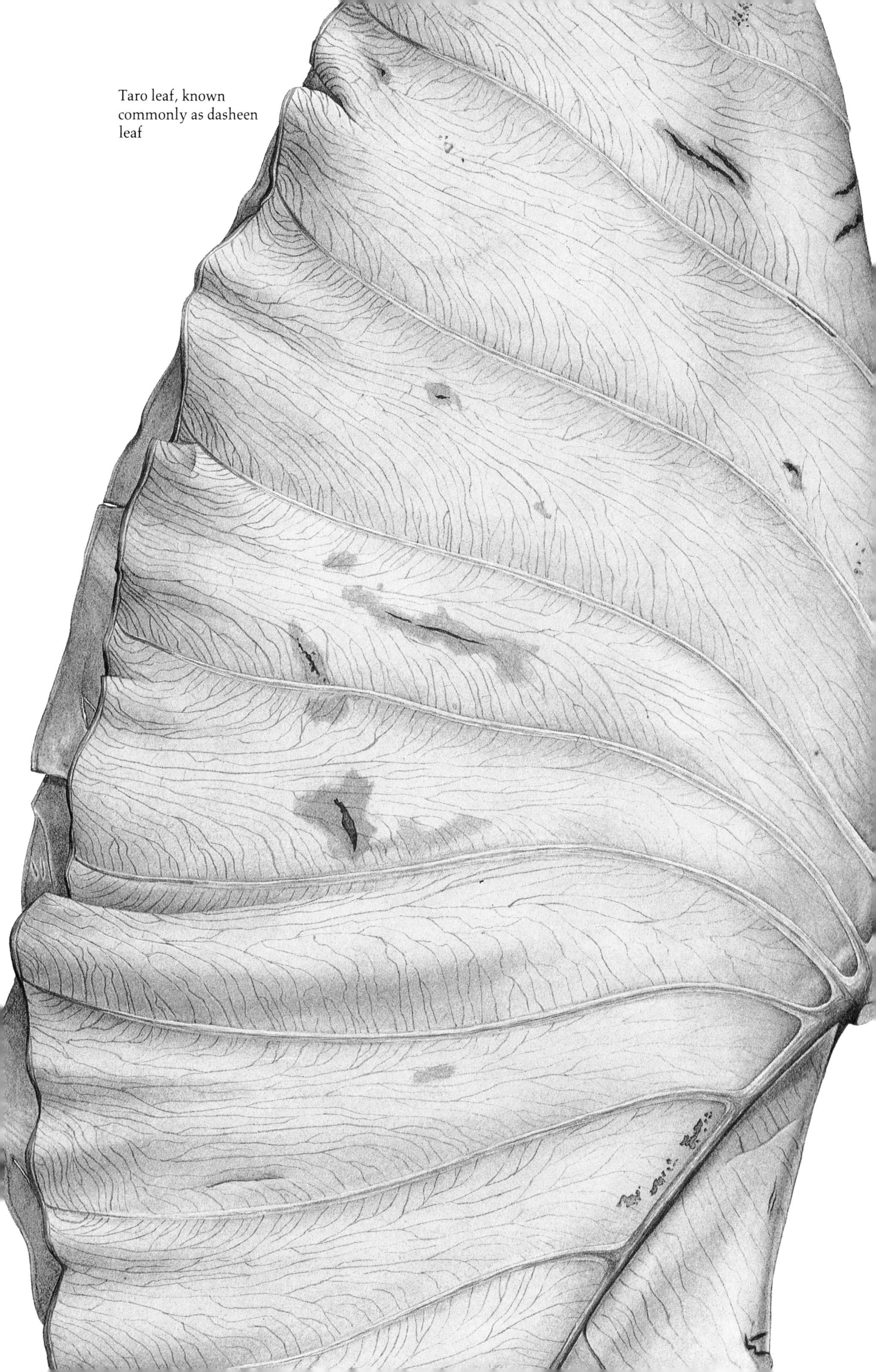

Taro leaf, known commonly as dasheen leaf

Taro and Onion (Arbi Pyaz)

In this recipe of Meera Taneja's, taro is simmered first then fried to crispness, mango powder giving an agreeable sharpness.

Boil 500 g (1 lb) taro until it is just cooked. Peel and cut into long thick strips. In a heavy pan heat 30 g (1 oz) ghee and fry 1 tsp cumin seeds, and ¼ tsp each mustard and *kalonji* (nigella) seeds until they pop and splutter. This should happen almost immediately.

Mix in 2 medium sliced onions and keep them moving until they are brown. Put in ½ tsp hot chilli powder, 1 tsp *amchoor* (mango powder) and the pieces. Raise the heat and keep the pieces moving. Scrape the pan well as the vegetables begin to stick and turn them out on to a serving dish when they are nicely crisped. Sprinkle with 1 tsp garam masala.

Eddoe Soup

Here is a soup made from the tubers of taro (or related plants). Its stickiness gives a smooth texture to the soup, which is flavoured in rather similar ways to callaloo.

1 kg (2 lb) eddoes, peeled, sliced
125 g (4 oz) each fresh and salt pork, or gammon or green bacon, coarsely chopped
1 medium onion, chopped
large pinch thyme
4 spring onions, coarsely chopped, or small bunch chives, chopped
salt, pepper
knob butter and chopped parsley to finish

Cook eddoes, and meats, onion and thyme, together in 1¼ l (2¼ pts) water. Process briefly, in batches, so that the eddoes are crushed completely but the meat retains some identity. Alternatively pour into a *mouli-légumes* set over a bowl, pick out the larger pieces of meat and sieve the rest through, with the eddoes.

Return to the pan, add the spring onion or chives and season to taste. Simmer gently for 5–10 minutes, then stir in the butter and parsley.

Gaabhaa Nepali

In other words, elephant ear shoots as cooked in Nepal. Allow 375–500 g (¾–1 lb) for 4 people, as part of a course. You can just cut the shoots into convenient lengths, but the Nepali way is to peel off the fine outer skin and wash the leaves well. They are then rolled up again and the whole thing bent round and tied in lovers' knot style. Other ingredients:

½ tsp ground turmeric
¼ tsp ground coriander seed
¼ tsp ground cumin
2½ cm (1 in) piece ginger, peeled, chopped
¼ tsp hot chopped chilli or cayenne pepper
¼ tsp fenugreek seeds
2 tbsps mustard or sunflower oil
lemon juice or vinegar
¼ tsp celery seeds
1 tbsp ghee
salt

Crush the first 5 ingredients (apart from the shoots) to a paste. Fry the fenugreek briefly in the oil, then add the spice mixture and a little salt. After a minute put in the shoots. Keep stirring and cook for 5–10 minutes, adding a little water to make a sauce. Sharpen with a little lemon juice or vinegar. Meanwhile fry the celery seeds in the ghee, then add to the curry at the end. Serve rapidly.

Elephant Ear Shoots Western Style

These tender shoots are delicious enough to be eaten on their own, like asparagus or sea kale. Prepare them as above, only cutting them into pieces if they are too long to cook and eat comfortably otherwise. Cook them as lightly as possible, drain them well and serve them with one of the traditional sauces, hollandaise, melted butter, cream and butter sauce and so on.

Taro runners

Callaloo

Callaloo is the greenery of taro and related plants, and you may well come across it under many different spellings. It has given its name to this famous Caribbean soup, which is normally regarded as a speciality of Trinidad, though other islands make their own variations.

500 g (1 lb) shredded callaloo leaves
1 l (2 pts) chicken stock
125 g (4 oz) gammon, diced, or salt pork and bacon, chopped
1 onion, chopped small
1 small hot red chilli
pinch of thyme
2 cloves garlic, chopped small
3 spring onions, coarsely chopped, or small bunch chives
250 g (8 oz) shelled crab meat, prawns or large shrimp
juice of 1 lime
250 g (8 oz) okra, trimmed, sliced
coconut milk (see p. 39)
knob of butter
salt, pepper

Simmer first 8 ingredients together until the meat is tender. Remove the chilli when you find it hot enough for your taste. Mix shellfish and lime juice. Add the okra to the soup, and when that is almost cooked, put in the shellfish (keeping back any juices) and the coconut milk. Let the panful murmur on a low heat, below boiling point, for about 5 minutes. Stir in the butter, then adjust seasonings, adding the shellfish juices if any further sharpening is needed. Dilute with extra stock or water if you prefer a thinner consistency.

Serve with rice to stir in, or cooked plantain pounded and rolled into little balls.

Eddoe (Taro) Fritters and Mexican Red Sauce

You can make fritters of any of the root vegetables by peeling them and grating them into some kind of pancake batter. This is most successful for sweet potato, which can then be served with honey or butter and sugar. Other roots demand a savoury treatment – a Mexican red sauce goes well with them (see below). Thicker fritters of grated, starchy tubers need no flour: peel and grate coarsely 500 g (1 lb) eddoe. Add the peel, finely grated, of a small lime, 1 tbsp sugar, 60 g (2 oz) cream cheese and finally an egg, large size. Beat well. Using 2 tablespoons drop the mixture into a pan of hot deep fat. The three Browns, from whose *South American Cook Book* this and the following recipe come, say that these fritters (which they make with *yautia*, a similar type of tuber) should be put round meat.

They could also be served with Mexican red sauce, a delicious peppery blend, thickened with nuts, which makes a welcome change from our familiar forms of tomato sauce.

1 onion, chopped
3 tbsps olive oil
2 garlic cloves, finely chopped
4 tomatoes, skinned, seeded, chopped
4 red peppers, seeded, chopped
125 g (4 oz) ham, finely chopped or minced
2–4 red hot chillies, crushed
2 tbsps sesame seeds
1 tbsp ground blanched almonds
1 tbsp ground peanuts
1 clove
pinch ground cinnamon
pinch ground ginger
pinch ground mace
1 tbsp sugar
½ tsp salt

Soften onion in the oil without colouring it. Put in the garlic and continue the slow cooking until it has blended in well. Put in tomatoes, peppers, ham and chillies. Simmer until very thick.

Roast the sesame in a dry non-stick frying pan, until it starts to pop and jump. Grind, electrically if possible, with almonds and peanuts. Add to the pan with remaining ingredients. Simmer for about 5 minutes to blend well. Store, covered, in the refrigerator.

This thick blend is more of a paste to spread on things than a sauce. To serve it with vegetables, meat and poultry, thin it with hot water or stock.

Pepperpot Soup

A Jamaican dish, another wonderful soup, based on Elisabeth Lambert Ortiz's recipe from her *Best of Caribbean Cooking* (she also gives a number of stews using yams, *yautia*, eddoes and sweet potato). It uses pretty well all of the plant, since the petioles and young shoots could be put in under the heading of callaloo just as well as the big elephant-ear leaves. The smoothness of the soup comes from the tubers as well as from the mucilaginous okra and gelatinous shin of beef. Dumplings, of the English kind, can be added towards the end.

500 g (1 lb) shin of beef, cubed
250 g (8 oz) salt beef or salt pork, cubed
500 g (1 lb) kale, chopped
500 g (1 lb) callaloo, chopped
1 onion, chopped
2 cloves garlic, chopped small
4 spring onions, chopped
½ tsp dried thyme
1 green hot chilli, seeded, chopped
250 g (8 oz) yam, peeled and sliced
250 g (8 oz) eddoe, coco or taro, peeled and sliced
salt, black pepper
12 okra, sliced
a little unsalted butter
125 g (4 oz) shelled cooked shrimps or prawns
250 ml (10 fl oz) coconut milk (see p. 39)

Put the meats into a large pan with 1¼ l (2¼ pts) water and bring gently to the boil. Skim well, and simmer, covered, for about an hour. Put the greenery into another pan with ½ l (¾ pt) water and cook until tender. Then sieve, blend or process to a purée. Tip into the meat pan with all the ingredients down to and including the eddoe, coco or taro. When the meat and root vegetables are tender, taste and add seasoning. Leave to burble gently while you cook the okra in a very little butter until lightly browned. Put into the soup with the shrimp and half the coconut milk. Heat through, below boiling point, for 5–10 minutes then taste again and add more seasoning and coconut milk, if you like.

Yahni with Colocassi and Celery

Cypriot colocassi is easy to recognize from the characteristic stump. Try to buy one piece of the right weight, as different pieces can cook at different speeds. Wash and scrub it well, then peel it without more water, like a potato, and remove the stump. Leave it whole.

The great point of Greek and Cypriot stews is to cook them so that the liquid reduces to a rich thick sauce by the time the meat is tender, a technique more suited to the dying heat of charcoal or an old-fashioned oven than to our gas and electric hobs. Such dishes always seem to cook better in earthenware than metal: be prepared to slip an asbestos mat over a gas burner.

1½ kg (3 lbs) pork loin or shoulder or a chicken
groundnut or sunflower oil
1 large onion, sliced
½ l (¾ pt) tomato juice or sieved tomatoes
head of celery, prepared and cut in 1½ cm (½ in) pieces
1½ kg (3 lbs) colocassi, peeled
juice of 1 lemon
salt, pepper

Cut the pork into serving pieces or large cubes, or joint the chicken. Brown in a little oil with the onion. Pour off surplus fat. Add the tomato and enough water to cover. Stir in the celery. Chip off pieces of colocassi into the pot and sprinkle them with some of the lemon juice: it is prudent to keep them on top, as they need less cooking than the pork or chicken legs. Cover and cook gently until tender. Keep an eye on the pot, and when the meat is almost done remove the lid and allow the liquor to bubble down. Season at the end with extra lemon juice if you like, and salt and pepper.

I should add that Greek cooks do not often manage these sort of stews well. They overcook the meat to shreds. It may not be the correct peasant style, but you will get a better result if you remove bits of meat as they become tender (this applies particularly to chicken breast) and the colocassi. Then you can raise the heat and boil down the sauce without damage.

Sweet-sour Patra

A dish that delighted me in a vegetarian meal we once ate in Manchester, at an Indian restaurant, was patra. It consisted of pinwheel slices of contrasting green and fawn which had been fried a nice brownish-red colour. We tried to understand the instructions given us, and set off next day for the splendid row of Indian and Pakistani shops down the Wilmslow Road. There we found elephant ears – taro leaves – as large as catalpa leaves, the *besan* and other items.

12 taro leaves, trimmed
100 g (3½ oz) tamarind
60 g (2 oz) gur *or brown sugar*
150 g (5 oz) gram flour (besan)
1 level tsp powdered turmeric
1 level tsp ground coriander
2 level tsps hot powdered chilli
1 level tsp garam masala
2 tbsps chopped coriander leaves
2 tbsps grated fresh coconut
wedges of lime or lemon

Wash, dry and divide the leaves into 2 piles, smooth side down. Pour boiling water over the tamarind to cover. Cool then sieve the pulp and discard seeds and fibre. To the sieved pulp add the sugar, flour and spices, with enough water to make a thick but spreadable paste.

Take the first leaf, spread it with the paste. Put the next leaf on top. When you finish the first pile of 6 leaves, do the second one. Roll each pile up and tie with button thread or thin string. Steam the 2 rolls for 30–45 minutes, until they are tender when pierced with a skewer. Remove, cover with a cloth and cool to tepid. Cut into slices, about 1 cm (½ in) thick, and deep-fry to a reddish brown.

Serve scattered with the coriander and coconut, with the lemon or lime wedges tucked in.

Taro shoots

WHITE RADISH

(Raphanus sativus var. longipinnatus*)*

You will not find radishes in the wild, only in gardens or on farms. The long white radish in the illustration is a cultivar – in other words a horticultural variety that was begun and continued under cultivation, becoming so important that it needed a name of its own. Which is to say that we have been eating radishes, or rather several cultivars of radishes, for a very long time.

The white winter radish comes newly to Britain from Asia and the Far East with a gentle juicy crispness that blends tactfully with many salads, more tactfully than the little round red radish of summertime, which can be quite fierce. But the possibility of fierceness is there: one feels the subduing hand of a succession of gardeners, just as one does with the bite of celery or the bitterness of cucumber. In the seventeenth century there was still the element of luck about such things, in Japan as in Italy or France or England.

In the bitter radish
That bites into me, I feel
The autumn wind.

That was Bashô, contemporary of John Evelyn, who also knew about the mordacity of radishes, knew they should 'eat short and quick, without stringiness, and not too biting. These are eaten alone with salt only, as carrying their pepper in them.'

In fact I think we owe the introduction of this good new root vegetable to Asian immigrants, rather than to the Japanese, since it appears in our shops under various spellings of the Hindi name, *mūli*. On the whole, though, I prefer the Japanese style of exploiting this long radish as a lightener to winter stuffiness. If it is cooked it becomes a gentle relation of the turnip, little more, and its extraordinary virtue of seeming always chilled, always cool to the mouth and crisp, vanishes. And it is through coolness and crispness, I suspect, that it will make a niche for itself in our Western diet.

To Choose and Prepare

Avoid roots that begin to feel a little soft and bendy: the firmer the better. Trim off the top and the end, then cut in sections as required and peel.

Chrysanthemums

Cut off a length of white radish, or daikon as it is often called, and trim it into a rough globe. Place it between 2 chopsticks, then cut down at right angles to the chopsticks into thin slices – the chopsticks prevent the knife going right down. Turn the piece of daikon and cut the other way. Soak in salted water for at least 30 minutes, then press gently on the cuts to spread them and make a flower effect. The chrysanthemum can now be soaked in a sweet pickle of 2 tbsps sugar and 1 tbsp rice wine vinegar, if it is to decorate smoked or plain cold poultry – duck, for example (white radish and orange salad also goes well with duck).

White Radish and Orange Salad

15 cm (6 in) white radish, peeled
4 oranges
salt
lemon juice
black pepper
18 black olives, stoned

Slice the white radish thinly. With a zester, remove thin shreds of orange peel to garnish the salad, then peel and slice the oranges, removing the pips. Season the slices with salt, lemon juice and black pepper. Make a ring of orange slices on a round serving plate, then an inner circle of white radish slices. Put the black olives in the centre and top with shreds of orange peel.

Good with smoked or salted duck, or with ham, or on its own as a first course.

Sleet

The name comes from the look of shredded daikon. Peel a couple of equal lengths and grate them in the processor; or grate them coarsely by hand; or else cut sections of daikon as if you were peeling them continuously so that you get a long sheet, or sheets – roll them up and slice across into shreds. Mix with a little lemon juice or vinegar. Season as appropriate, with salt and a pinch of sugar.

Sleet makes a good contrasting accompaniment to gravadlax and mixes well with golden whitefish caviare or red salmon caviare – roughly equal quantities; or to make a very rustic salad, with grated carrot; or else with cooked sliced mushrooms (the Japanese use a particular kind called *nameko*).

Parsley can be used to decorate such salads, as can quarter slices of lemon or shreds of lemon or lime peel. I nip the green shoots from my strings of onions that always seem to sprout round about the New Year and slit them in two: half the shreds I put in iced water to make them curl, the rest stay plain and straight, a simple but elegant decoration. Spring-onion flowers also go well with daikon salads – cut the spring onions into lengths and cut one end down for a while and the shreds will spread out in a flowery way.

Sashimi

¾kg (1½lbs) sea bass, sea bream, sole or plaice
250g (8oz) slice pink tunny or 375g (12oz) mackerel
7½cm (3in) white radish, shredded, or equivalent quantity sleet salad (see above)
shoyu *(Japanese soy sauce)*
lime juice (optional)
1 level tbsp wasabi *(green horse-radish powder) or 2 level tbsps finely grated fresh ginger*
3 Welsh or spring onions

Fillet the fish as necessary, leaving the skin. Wrap in greaseproof or muslin and chill in the freezer for an hour or two to firm up.

Pour *shoyu* into 6 little bowls to make a dipping sauce, or use half *shoyu*, half lime juice. 6 tbsps in all should be enough if your bowls are tiny.

To assemble: no more than half an hour before the meal, mix the *wasabi* to a thick paste with a little water. Let it stand for 20 minutes or so. It should then be thick enough to hold its shape. If you use ginger instead, shape it into 6 little mounds.

Then slice the chilled fish paper thin, on the diagonal (best for people who are unfamiliar with *sashimi*), or into ½cm (¼in) slices, or – with flat fish – into strips. Discard the skins, pick out any tiny bones with tweezers.

Arrange on 6 chilled plates, in, for example, a sunflower effect, with the darker slices in the middle: put a blob of *wasabi* in the centre, or a mound of ginger. Put the radish or sleet salad in 6 bowls slightly larger than the *shoyu* sauce bowls.

Cut the onions into slices or curled shreds as garnish.

To eat: mix the *wasabi* or ginger into the *shoyu* according to taste, then some of the radish or sleet salad. Pick up a slice of fish and dip it into this mixture, using chopsticks or a fork.

Rettischaum (Radish Foam)

Process or grate 10cm (4in) radish into julienne strips. Sprinkle with salt, leave at least 30 minutes, then drain and squeeze out moisture. Whip 150ml (5floz) double cream until stiff, fold in the radish shreds and season to taste.

Add ½ tsp caraway seeds or mix in 30g (1oz) chopped walnuts.

Serve as a sauce with cold meats, beef especially, and pastrami, smoked poultry and game, or with fish. An excellent sauce for those who find horseradish too pungent.

By using more radish, or mixing in cooked beans, you can turn this sauce into a salad.

J.L.M.

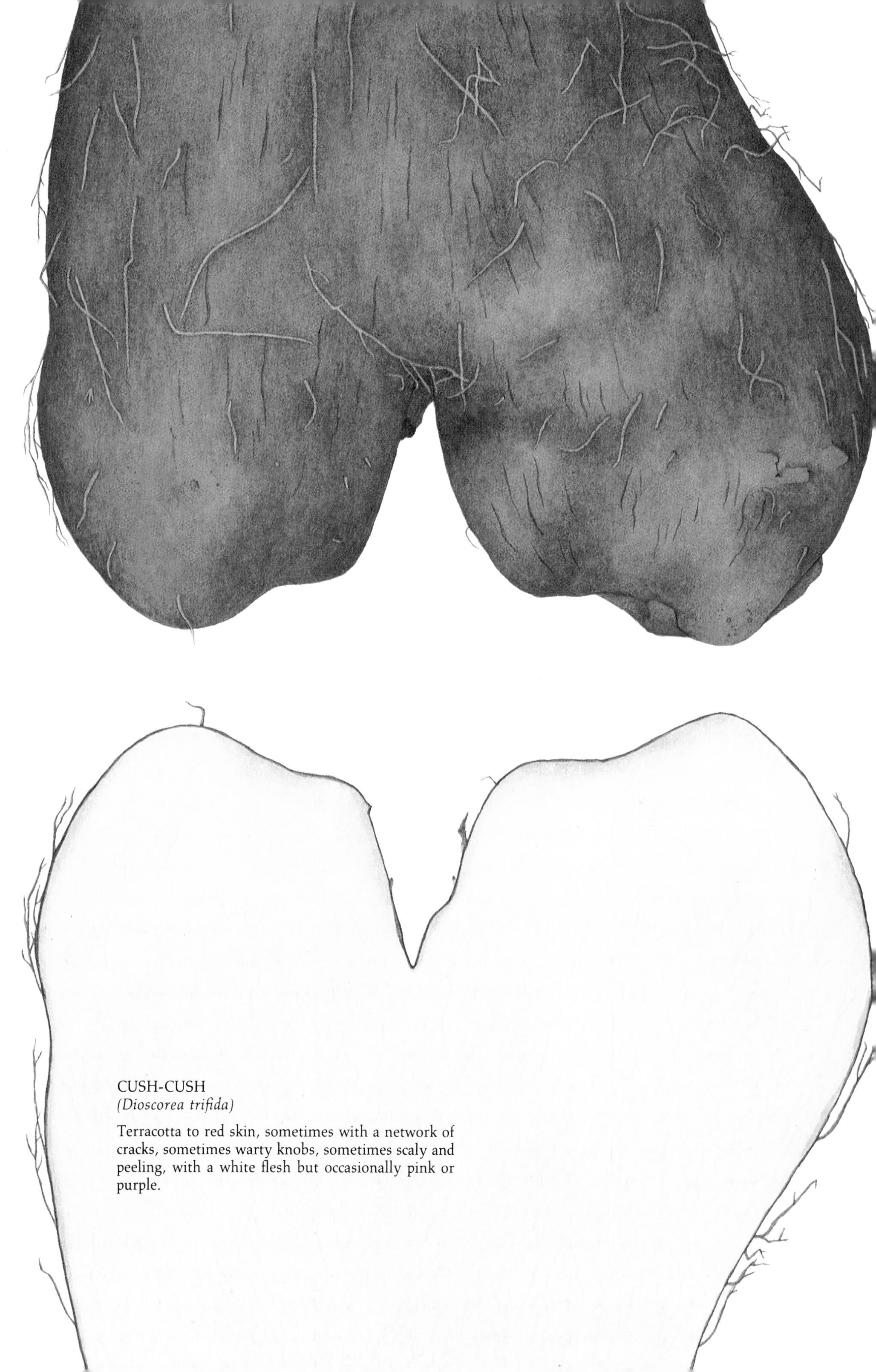

CUSH-CUSH
(Dioscorea trifida)

Terracotta to red skin, sometimes with a network of cracks, sometimes warty knobs, sometimes scaly and peeling, with a white flesh but occasionally pink or purple.

YAM

(Dioscorea genus*)*

Yams are not sweet potatoes (though you will find that sweet potatoes are called Louisiana yams in the southern states of America and such names as *nyam* and *iyan* in West Africa).

Yams are not taro either (though in some parts it is called cocoyam).

Moreover a yam bean is not the dependent pod of a true yam, but another, sweeter, crunchier tuber, also called *jicama* (in Mexico), and *Pachyrrhizus erosus* by botanists.

True yams all belong to the genus *Dioscorea*, named after the first-century botanist and physician, Dioscorides. I am not sure why. Not a very complimentary gesture, since most of the 600 species in the genus are poisonous, and the edible species – about ten – are dull. Very dull indeed, I would say. Perhaps I have been unlucky. Yam-on-the-spot may be a delight, but by the time it gets to markets here it is a bore. One understands the vast success of the potato. Of course for a staple food you do not want the positive kind of taste that characterizes, say, asparagus (this is why the Jerusalem artichoke, which looked in the seventeenth century as if it might be Europe's answer to famine, so obligingly does it ramp through gardens and fields, was eventually relegated to cattle fodder and an occasional winter soup or salad for the discerning eater). Even given the mild flexibility required, yams do seem to me to be unnecessarily bland. Africans and West Indians of African descent partner it with hot pepper sauces and stews spiked with salty dried fish. Latin Americans stuff it into their vast sociable stews, along with sweet potato and taro and a host of other items, to sop up the sauce and fill the cavernous stomachs of huge and hearty eaters.

Yams come in several shapes, some like the white yam very regular and cylindrical. Others, the large-fingered yam or water-yam for instance, the cush-cush and the sweet yam, grow in all manner of lumpy forms. They have the look of full bags, an udder-like shape sometimes, heavy yet soft-looking, that suggests to me a host of Polynesian and African folk tales that have no contact at all with the very human mythology of our Mediterranean culture.

What is interesting about yams is their name, the word yam itself. Sometimes it means little more than 'starchy tuber', covering whatever happens to be the predominant tuber in a country's diet, whether it be the taro or the sweet potato. It is as if the name described a function rather than an identity.

The first trace of the word in Europe is in Portugal. The story of its arrival there, an extraordinary one, was disentangled by Isaac Henry Burkill (1870–1965), the great scholar and author of *A Dictionary of the Economic Products of the Malay Peninsula*. In a paper delivered to the Linnaean Society in 1938 he began by considering another of the world's great food plants, taro (pp. 80–9), pointing out that the Portuguese name for it, *curcas*, 'supplies the last of the links of the chain – "kolokasion" – "elkulkas" (Arabic) – "alkolkaz" (Spanish) – "curcas" – which are the footmarks on the track of the dissemination of the crop along the southern side of the Mediterranean'.

In its African journeys, the Greek word *kolokasion* underwent even more adaptation ending up, unrecognizably to the ordinary person, as *koko* on the extreme west coast. That was the name carried by slaves to the West Indies and then back again to Liberia. And so we get taro as cocoyam.

The journey to Portugal of the word yam, if not its reality as a tuber, is also a matter of slaves – fifteenth-century slaves this time.

Portugal nowadays means to many of us a thankfully peaceful and backward country, a holiday haven with enough relics of a great past to intrigue the more cerebral tourist whose pleasures are not totally enclosed in the triangle of sun, sand and grilled sardines. In the fifteenth century it was a very different place, with what some politicians approvingly characterize today as a 'go-go economy'. The Portuguese were go-going down the coast of Africa. In 1415 they relieved the Moroccans of a port or two and managed to clear the seas down the coast, making them safe for exploration and exploitation.

In 1441 'a scrimmage at the Rio d'Oro put some Moorish prisoners into Portuguese hands. These were ransomed by a payment in gold and the substitution of negro slaves. The arrival of these slaves in Lisbon . . . created great excitement, and the wretches were sent as curiosities to the Pope in Rome. They pass out of this history, but left in Lisbon a desire for more slaves.' The possibilities of this new trade were so unlimited that soon slaves were working the land and reclaiming the Moors' old rice fields in the south, liberating 'the limited Portuguese manpower for war and adventure . . . in 1535 a resident Dutchman – a teacher of Latin, Nicolaus Cleinard – wrote somewhat vulgarly that the country was filled right to belching with slaves.'

These slaves had been enslaved already; they were the Temne people of the West African coast at Sierra Leone and present-day French Gambia, subject to a bossy and invading tribe, the Mandingo. The Mandingo belonged to the Middle Niger. They had been traders in gold, taking it to the coast, bringing back salt. When they came under pressure from the Sudan they gradually moved west permanently, taking with them their skill in rice-growing, the cruel habit of slavery they had learned from the Arabs and the word *nyamba* for the large yam, *Dioscorea alata*.

Soon the Mandingo had the indigenous Temne growing rice for them, and no doubt yams as well. When the Portuguese arrived the Mandingo were able to sell them just the right kind of trained labour for reviving rice culture in southern Portugal. Burkill suggests that it was these poor Temne people who took their name for *Dioscorea cayenensis*, *'enyame'*, to Portugal and applied it to the tuber they found established there, the taro.

Burkill lists the related names used in different African languages. He finds that they also covered sweet potatoes and taro, and cassava root as well. Back we are with 'starchy tuber', all the untidy nomenclature of the market places which botanists try in vain to clear up and restrict. To past cooks – whether they used *inhame* (Portuguese), *ñame* (Spanish), *igname* (French) or yam – the one word was practical, meaning that recipes were interchangeable and the use if not the taste of the vegetables was the same. To the more anxious and knowledgeable cook of today, possessor of the cuisines of many countries whether by way of the bookshelf or the supermarket, the one word means confusion and panic. I would only say that such apparent verbal sloppiness, wherever it occurs, was practical in less bookish times, and, if the reasons are understood, can be helpful today.

To Choose

Although, as I have implied in the introduction, all yams are treated in roughly the same ways, there are subtle differences that are useful to know about if you are faced with a choice. In general terms, of course, buy yams that look in good fettle, not too bashed about and bruised (the sweet yam bruises easily). Always allow plenty of time for cooking: some yams are harder than others, for instance the white yam.

To Prepare

All yams except for the sweet yam *(D. esculenta)* contain dioscorine, which is poisonous. Cooking destroys it.

Scrub them well, then cut them up according to the recipe. As they show a tendency to discolour – i.e. if they have to wait around – drop the pieces into salted water.

Baked Yam

You will get the best results by baking large pieces of yam, with their peel on, over a

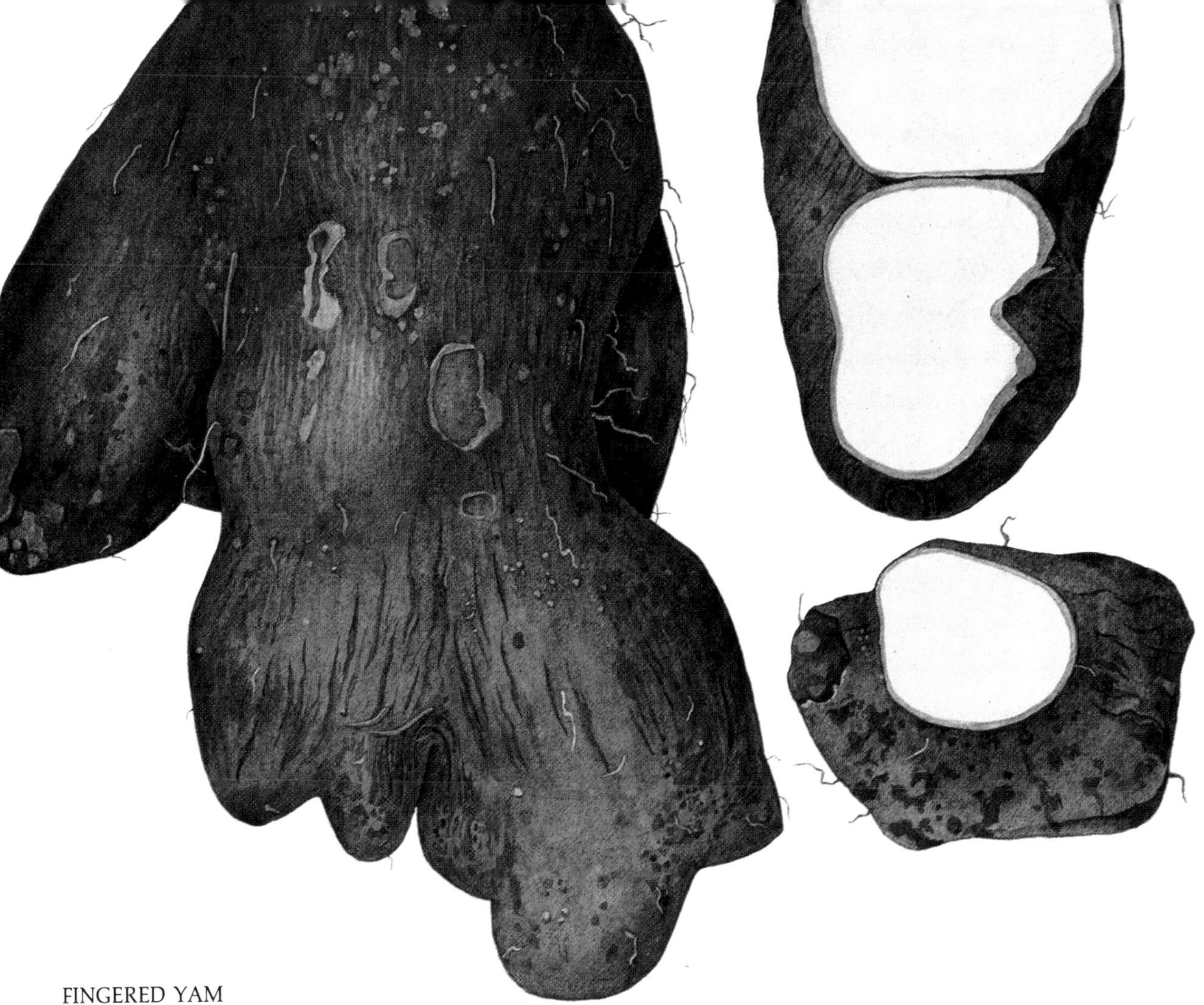

FINGERED YAM
(Dioscorea alata)

A cultivar of the greater Asiatic yam, with finger-like protuberances, dark brown skin and white flesh sometimes tinged with pink.

barbecue fire. It gives them an agreeable smokiness. They are done when a skewer goes in easily. Baste them with a fat appropriate to the rest of the meal, and sprinkle them finally with salt. They can also be baked in the oven like potatoes, but are dull by comparison. Push a skewer through the middle to help conduct the heat, and mash up the cooked flesh with plenty of butter and seasonings, for instance a splash of Angostura bitters.

Pounded Yam (Fufu)

Boil peeled yam in water until tender. Starting with one large piece, pound it in a mortar, adding more and more pieces as the mash starts to leave the sides. You can add a little boiling water to make the job easier. Or you can have recourse to a processor or very powerful mixer. Aim to end up with a silky dough that is soft but holds its shape, rather firmer than mashed potato but not as coherent as polenta. Add salt to taste. Put on to a hot serving dish and mould it with your lightly moistened hands. Warm through in the oven. Serve with spicy dishes, with stews and thick soups, e.g. okra soup (p. 71).

The mixture can be varied by starting off with equal quantities of yam and plantain or yam and sweet potato. Start cooking the yam pieces, then as they begin to soften put in the peeled sweet potato. The peeled plantain should go in when the yam is nearly tender.

Boiled Yam

Peel after cooking in salted water. Can be finished in butter with plenty of pepper, unless it is to be served with a piquant sauce or stew.

Sprats and Pepper Sauce (Yoyo Stew)

An African dish that can be made in Britain with some hope of authenticity, since sprats are the fish and the seasonings are easily available. Our tomatoes being so bland, add tomato purée and a pinch of sugar to give some zip.

500 g (1 lb) sprats
salt, pepper, mixed spice

sauce
375 g (12 oz) tomatoes, skinned, chopped
1 medium onion, sliced
1 sweet red pepper, seeds discarded
1 small red chilli, seeds discarded, or 1 level tsp hot chilli powder
1½ level tsps salt
pinch sugar
1–2 tbsps tomato paste or purée
groundnut oil
pinch thyme and ground coriander seed (optional)

Rinse the sprats – no need to gut them – and dry them with kitchen paper. Sprinkle with salt, pepper and a little mixed spice. Set aside while you start the sauce.

Pound in a mortar, or process, or blend the tomato, onion, pepper, chilli and salt, with the sugar and half the tomato paste or purée.

Heat enough oil to cover the base of a sauté pan. Fry the sprats briefly on both sides until they are lightly browned, remove them and keep them warm.

If the oil is murky, strain it off into a clean pan. Otherwise continue in the sprat pan.

SWEET YAM
(Dioscorea esculenta)

A cultivar of the lesser Asiatic yam, ovoid or sausage shaped, generally no more than 20 cm (8 in) long, with a thin fawn to brown skin which bruises easily and small feeding roots.

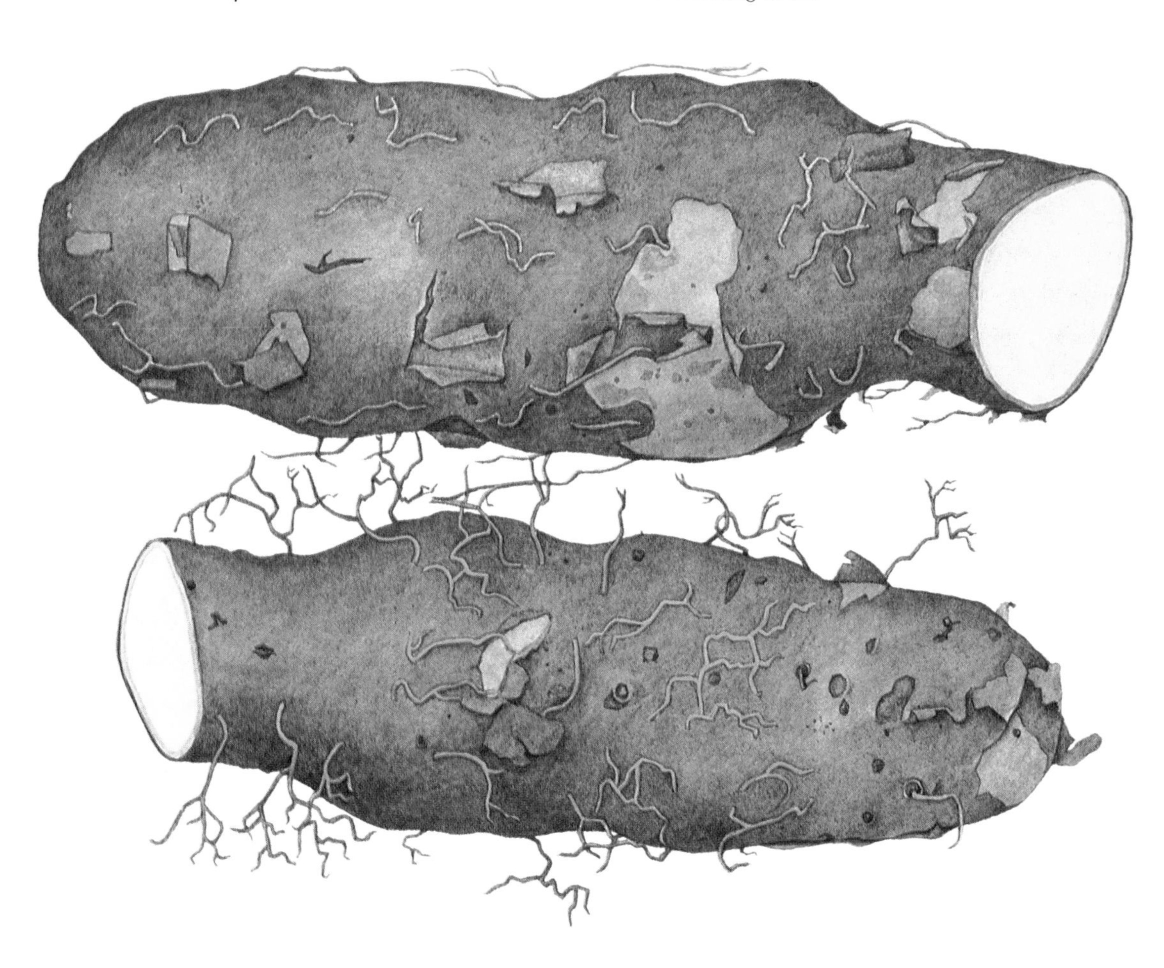

Fry the tomato and pepper mixture until it is almost dry, stirring it all the time. The wateriness of the tomato should evaporate. Taste and check the seasoning, add more tomato purée, and add the thyme and coriander if you like.

Mix the sprats in carefully and turn on to a hot dish. Serve with pounded or boiled yam, or with plantain.

Note: you can adapt the recipe to fresh fish and prawns. Make the sauce as above but use 4 sweet peppers and 1 tomato. When it is nearly dry, put in the fish to simmer, with an extra tomato, sliced, and finally the shelled prawns. Monkfish is a good choice as it does not break up too easily.

Wateryam Fritters *(Ojojo)*

Fritters of the moist wateryam are popular all over Africa, and are sold as street snacks. If the mixture becomes too wet, add a little ground rice.

Peel 500 g (1 lb) of wateryam and grate it finely. Mix in a small chopped fresh red chilli (remove the seeds first), and a small chopped onion. You can also add just 20 g (2⁄3 oz) of chopped tomato. Salt the mixture to taste. Beat well to lighten the mixture. Fry spoonsful in hot oil.

Thick Goat Meat Pottage

Some of the ingredients, the flavourings, in this recipe you will only be able to find in shops that cater for West Africans. Potash helps to prevent palm oil separating from the sauce and flavours the food. The best palm oil is flavoured with various spices and coconut and is difficult to find even in Nigeria. Good market palm oil is orange-red, and may be oily or almost solid. 'It can be eaten without cooking, can be used directly without frying in stews and will not give an unpleasant odour or taste to the product.' African nutmeg *(Monodora myristica)*, also known as the Calabash nutmeg, and *enge (Xylopia aethiopica)* or Guinea pepper are two other popular seasonings in tropical Africa. For more details turn to the *Nigerian Cookbook*, by H. O. Anthonio and M. Isoun from which this and the next recipe come.

500 g (1 lb) goat meat, kid or pork
1½ level tsps salt
pinch potash
1 enge *pod, crushed*
2 seeds African nutmeg, ground
1 level tsp dry red chilli, ground
1 small onion, chopped
2 medium green plantains, each cut into 3, peeled
500 g (1 lb) yam, wateryam or taro, peeled, cut up
2 tbsps palm oil
40 g (1⅓ oz) ground dry crayfish

Cut the meat into large pieces. Put into a pot with enough water to cover, with the next 6 ingredients. Simmer for half an hour. Add the plantain, yam, wateryam or taro, the palm oil and ground crayfish. Cover and simmer until the vegetables are tender. Remove the first pieces of vegetable to become tender, but allow enough to dissolve into the juices to thicken the soup. Should the meat be tender before this happens, just mash up some of the vegetables. Check the seasoning. The pottage can be diluted.

The meat can be removed from the pottage and served separately, or it can all go into one dish. Provide a salad as well.

Simple Yam Pottage

A Yoruba version of the famous African pepper soup which is not impossible to make even if there is no African store in your district.

Peel and cut up 1 kg (2 lbs) yam into large pieces. Cover it with water generously, add 1½ level tsps salt and 5 tbsps palm or groundnut oil. Pound or process or liquidize 150 g (5 oz) onion, 250 g (8 oz) tomato, 4 red seeded chilli peppers weighing about 15 g (½ oz) in all, and 40 g (1⅓ oz) dry crayfish. Add to the pan.

As in the recipe above, the smaller pieces should dissolve and thicken the mixture, which is more of a purée or stew than a soup (though you can always add more water, if soup is what you prefer).

It is a good idea to leave the pottage for a

while so that the flavours have a chance to develop. It can then be reheated and served with baked or fried fish, with meat and – if you are in Nigeria – with the vast land snails that weigh 300 g (10 oz) or more.

Caribbean Oxtail Stew (Sancocho de Rabo de Vaca)

This stew contains cush-cush yam which is the tastiest – though perhaps that is hardly the word – and the only one coming from America. Cassava root and *calabaza* or West Indian green pumpkin will come from the same shops and markets as the yam and plantain. As will the *melegueta* pepper, alias Guinea pepper, since it is native to that part of Africa bordering the Gulf of Guinea: its other name, grains of paradise, is perhaps a reference to its use in flavouring hippocras, though I would think it more probable it comes from the idea that the origin of spices was some earthly paradise (the Arab merchants who monopolized the trade kept their sources of supply secret, and fostered legends about them to put off the danger of rivalry in the business).

1½ kg (3 lbs) large oxtail pieces
4 cloves garlic, crushed with salt
2 medium onions
1 green pepper, seeded and chopped
1 level tsp each chopped parsley and coriander
3 grains melegueta *pepper*
1 level tsp origano
2 tbsps each cider vinegar and Seville orange juice or 1 tbsp each lemon and sweet orange juice
2½ l (4 pts) beef stock or half stock/half water
½ kg (1 lb) each white and yellow tannia *(yautia) or use taro, peeled and cut in thick 2½ cm (1 in) slices*
500 g (1 lb) cush-cush yams, peeled and cut in thick slices
500 g (1 lb) calabaza, *peeled and cubed*
500 g (1 lb) cassava root, peeled and cut in 1 cm (½ in) slices
3 ripe plantains, peeled and cut in 2½ cm (1 in) slices
salt, Tabasco sauce to taste

Put the oxtail into a huge casserole. Add all the ingredients down to and including stock or stock and water. Cover and simmer gently for 2 hours. This can be done in advance, which has the advantage that the fat from the oxtail will rise and set so that it can be removed easily. Bring back to simmering point, and add the remaining ingredients. Cook for an hour or until the oxtail meat begins to part easily from the bone. Taste and adjust seasoning, skim off any fat and leave until the meat is almost falling from the bone.

Serves 8.

Flying Fish Pie

I have occasionally been able to buy flying fish from enterprising fishmongers, but if you cannot find them fillets of firm white fish will do instead. In my experience it is not the flavour or texture of flying fish that is so remarkable but their beautiful shape with the wing-like fins that enable them to leap from the sea. It seems a shame to eat such exquisite creatures.

Season ¾ kg (1½ lbs) filleted fish and fry it lightly on both sides in butter. Then cut the pieces in two. Peel and cook 1 kg (2 lbs) yams, then cool and slice thinly. Slice a large onion thinly, and a large tomato, and 2 hardboiled eggs. Beat together 2 egg yolks, 2 tbsps groundnut or sunflower oil, 2 tbsps unsalted melted butter, 1 tbsp Worcestershire sauce and 6 tbsps dry sherry.

To assemble, butter a deep dish lavishly. Put in half the fish; scatter on top half the onion, tomato and egg. Then cover neatly with half the yams. Repeat and brush the top with melted butter. Pour over the egg-yolk mixture.

Bake in the oven preheated to gas mark 4, 180°C (350°F), for about half an hour until the top is brown and everything heated through. Be prepared to give it a little longer, but avoid overcooking at all costs.

Serves 4–6.

Elizabeth Raffald's Yam Pudding (1769)

'Take a middling white yam, and either boil or roast it, then pare off the skin and pound it very fine, with three-quarters of a pound of butter, half a pound of sugar, a little mace, cinnamon, and twelve eggs, leaving out half the whites, beat them with a little rose water. You may put in a little citron cut small, if you like it, and bake it nicely.'

Difficult to judge what Mrs Raffald meant by a 'middling' white yam, since they can be so enormous. Judging by other root puddings of the time, potato for instance, I took it that she had a 1½ kg (3 lb) yam in mind. Eggs were much smaller in the past than they are today, judging by an egg boiler that belonged to my mother-in-law: in total weight I judge they would come to about 500 g (18 oz), the size of our small size 5 eggs today. Use a moderate oven.

This is the earliest recipe for yams I have been able to find in English cookery books. Indeed it seems to stand alone until specialized cookery books of this century. The recipe is very similar to potato puddings of the time, which are also sweet affairs – no doubt because the first potatoes to arrive in Britain were sweet potatoes. And also because the division of meals into savoury dishes and a dessert, one kind salted, the other sweetened, was far less pronounced than it is these days. Dinner parties consisted of two or more courses of many dishes – in the second course puddings of all kinds were put on the table with roast game and other meat. Eating patterns were flexible and sugar was in any case a luxury until the development of trade with the East and with the West Indies in the seventeenth century caused sugar prices to tumble by nearly two-thirds, 1s 6d at the beginning and 6d per lb by 1700.

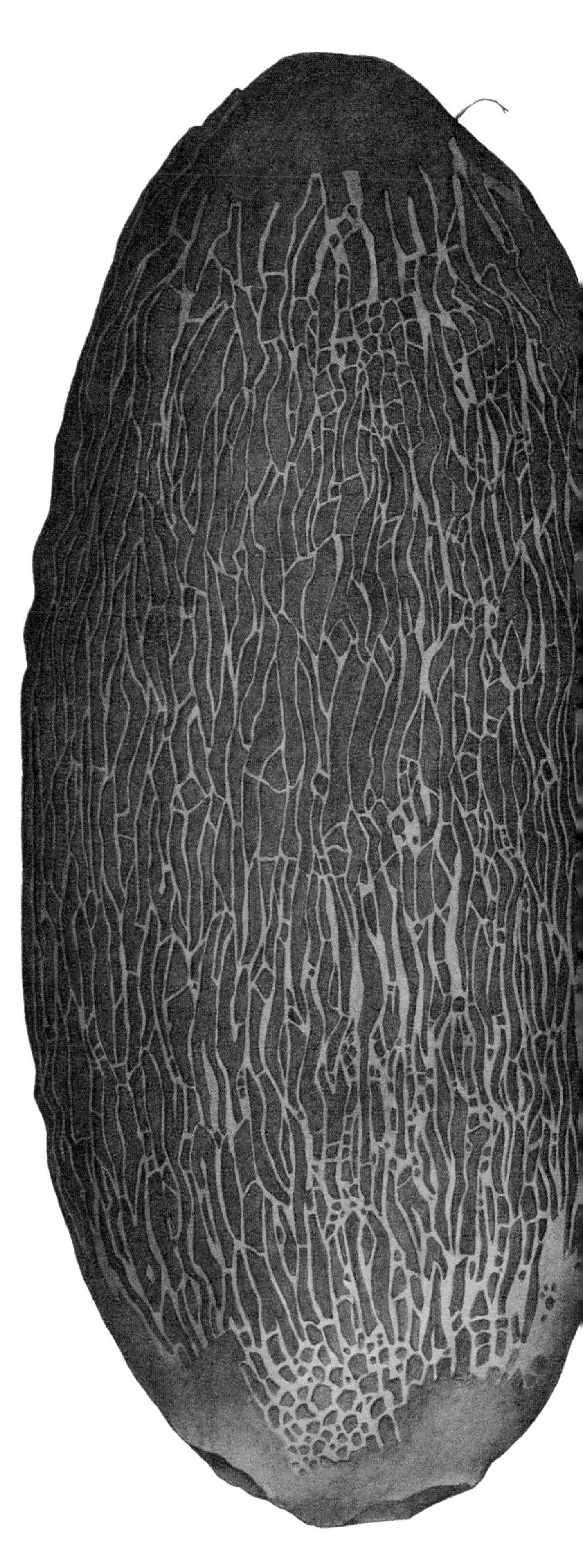

WHITE YAM
(Dioscorea rotundata)

Neat cylindrical shape, with brown skin, hard white flesh (takes longer to cook than other yams) and the one to use for African dishes as a general rule.

HERBS AND SPICES

BANANA LEAF

(Musa sp.*)*

The ribbed leaf of the banana (and some varieties are grown especially for the leaves) can be used as umbrella, roofing or plate, but for our purposes their role as a food wrapping is what counts.

The biggest wrap of all is for *kaua puaa*, the Hawaiian pork barbecue. Into a two-foot-deep hole go wood and stones. The wood is lit and reduces to red embers while also heating the stones. On top go banana leaves to make a steamy bed for the pig. Sweet potatoes and such like are pushed down to fill the gaps. Then another layer of banana leaves, sacks and earth. And 3½ hours later, not so long, the whole thing is opened up for the feast.

At the opposite end of the scale, a leaf may be cut into strips, midrib removed, warmed over a flame to make it more supple and then wound round small delicacies for steaming and frying. Try cutting a chicken into 12 pieces, marinating them in soy sauce flavoured with honey, mustard and vinegar, with coriander, garlic, sesame and ginger. Wrap them in an oblong of banana leaf, skewer with wooden toothpicks and deep-fry for about 8 minutes. The leaf gives an almost tea-like flavour, but you do not eat it. As with vine leaves, which give a faint lemonish taste, there is no true substitute for banana leaves, although from a purely mechanical standpoint aluminium foil works well.

In her *Indonesian Food and Cookery*, at the end, Sri Owen has two drawings of banana leaf in use – long cracker-shaped packages and a neat bowl in which cakes and puddings can be steamed, the sides where the leaf has been pleated up held firmly in shape with small wooden skewers. Her recipe for *nagasari*, rice cake with banana, consists of a batter made by beating 475 ml (16 fl oz) warm coconut milk mixed with 4 tbsps sugar into 250 g (8 oz) rice flour and 60 g (2 oz) cornflour. This is poured into the banana leaf cup; 2 large bananas cut each into half longways then across into three are submerged in the mixture and the whole thing is steamed for 60 minutes, cooled and eaten.

BETEL

(Piper betle)

While a brother-in-law of mine was Minister in Bastar in the Central Provinces, in the 1930s, the Maharanee died. With a little persuasion on his part, a tombstone was ordered from England, from Eric Gill. Eventually it arrived, a smooth light stone with chastely elegant lettering, and was set up in place. Soon, to my brother-in-law's artistic dismay, it was covered with trails and smudges of red. Faithful subjects had been kissing the stone while chewing *paan*, which are little packages of spices and so on wrapped in betel leaves; two of the items, *katechu* (from species of acacia) and quicklime, acting together, colour the saliva red and stain teeth.

The Central Provinces are famous for their betel gardens. In his *Dictionary of Economic Plants* (1889–96), G. Watt describes them and how carefully they are tended by a privileged brotherhood of betel growers. They are

> enclosed on all sides with a bamboo and mat covering to shield the delicate plant from the weather, and cool plantain leaves and the graceful wide-spreading leaves of the Arum, which shelters and supports the young plant, are massed within the walls. The interior of these gardens is strikingly pretty and inviting, the pan leaf carefully trellised in all directions – the broad leaves of the plant grouped beside it, affording a grateful shade, whilst the constant supply of water renders the gardens agreeably cool, even in the hottest weather.

The habit of chewing betel leaves, whether on their own or wrapped round spices and so on, was already an old one when Marco Polo described it in the thirteenth century. It was a habit then, an addiction and not just the final stage of a meal – which is how we are most likely to encounter it in the West. Often of course the betel leaf is not available, so beside the cash register in an Indian restaurant you may find glass bowls of some of the constituent spices, for customers to take a little pinch to chew on as they go out into the street, cloves perhaps, or cardamom, or fennel.

In India, as Madhur Jaffrey describes, 'a *paan* costs from a few pence up to £10 or £20, depending on what is wrapped inside it. At its simplest, this could be white lime paste, or *choona*; katechu paste, or *kattha*, a red paste made from the bark of a tree, which stains the mouth red; and chopped betel nuts' – these are nothing to do with betel leaf plants but nuts of the palm *Areca catechu*, round and hard, which need a special cutter with a V-shaped blade (I have one which is most attractive with handles in the form of scroll-like animals).

> Or the *paan* can contain combinations of fennel, cardamom, clove, perfumed nuts and spices, expensive tobaccos, drugs, and aphrodisiacs. An aphrodisiac like crushed pearls naturally pushes up the price a bit! The stuffing is placed on the centre of the leaf, and the leaf is then folded over in the shape of a flat triangle or a cone. Either shape is held in place by a clove, acting as a straight pin.

In London there is a shop in Drummond St, near Euston station, which sells twenty different kinds of *paan*.

CHINESE KEY

(Boesenbergia pandurata)

If you want to open a Chinese lock or soothe an elephant, this is the plant you need. Or rather you need the long brownish roots that hang from the corms. They are just the shape of a Chinese key, it seems – and this is the name the plant has been given in Malaya, Java and Thailand. The root is one of the items in an embrocation for sore muscles and joints. Human muscles I would say, though it is thought to work well with elephants, making them biddable and quiet (but how do you manage to approach the joint of an irritable elephant in the first place, I wonder?).

Mostly the roots are used as an aromatic spice and vegetable. As they give a hint of ginger they are used in fish curries. Thai cooks serve them raw with *khao chae*, a dish that gives its name to the whole course. It is much eaten in April, the hottest time, and consists of ice-cold rice flavoured with jasmine. By way of contrast various fritters and crunchy raw vegetables including Chinese keys accompany the rice.

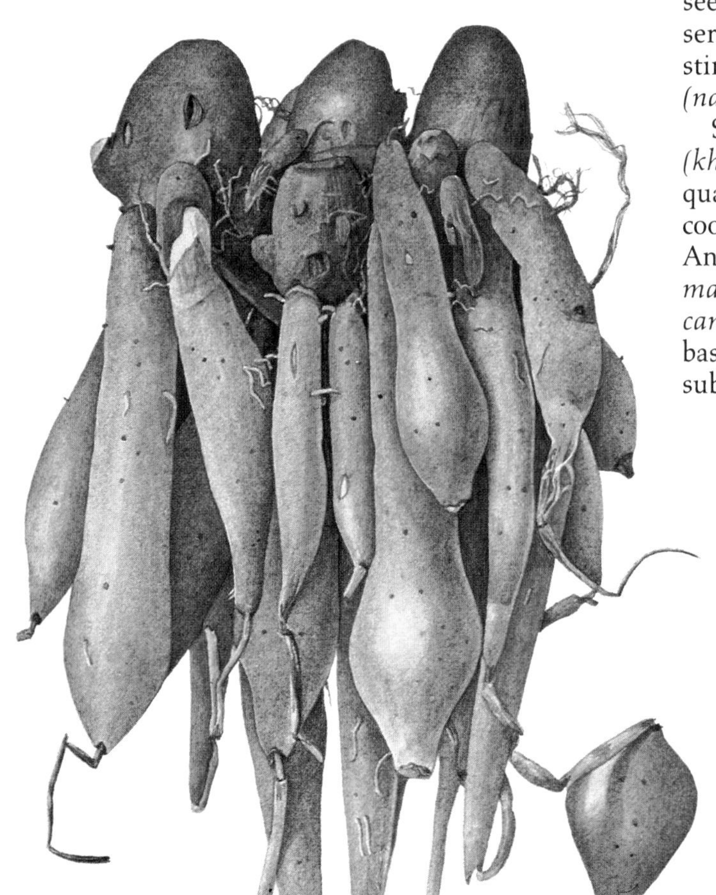

A simpler Thai dish for Western cooks is *nam ya pa*, a fish curry soup – more of a stew really. First you make 1½l (2¼pts) of coconut milk (see p. 39), taking off the top 175ml (6fl oz) of coconut 'cream' which rises. Then provide yourself with a plastic pot which once held 250ml (8fl oz) soft cheese or yoghurt. Into a pan pour the milk, plus

½ pot each garlic and shallot, both sliced downwards, and chopped lemon grass
1 pot finely shredded Chinese keys
2 tbsps shrimp paste (kapi)
1 tbsp fine shredded galangal
1 tbsp salt
½ pot flaked smoked fish, e.g. kipper, smokies, finnan haddock, without skin or bone

Simmer until everything is tender. Meanwhile shell 1½kg (3lbs) prawns or shrimp. Add to the soup, cook a minute or two to heat through and remove from the heat. Quickly add the coconut 'cream' and ½ pot of red and green chillies which have been seeded and sliced across, or you could use serrano chillies for a milder effect. Taste and stir in extra salt if required, or fish sauce *(nam pla)*.

Side dishes include boiled rice noodles *(khanom chine)* or boiled rice, hardboiled quail's eggs or quartered hen's eggs, lightly cooked bean sprouts and extra chillies. Another item on the Thai table would be *manglak* or hoary basil leaves (*Ocimum canum*, a species related to our familiar basil, *Ocimum basilicum*, which could be substituted, though with a different effect).

CURRY LEAF

(Murraya koenigii)

In the Corbett National Park, in Uttar Pradesh, Tom Stobart describes how the smell of curry leaves is 'overpowering as your elephant bursts through thickets of this plant'. As the leaves wilt the curry smell fades, so be careful when buying them and sniff thoroughly first. They are used in Indian and south-east Asian cookery, but more especially in southern Indian dishes, for example Madras curries, which are often flavoured with a large pinch of mustard seeds and a few curry leaves fried in ghee or oil. See sesame rice and avial (p. 121).

FLOWERING CHINESE CHIVE

(Allium tuberosum)

I am often grateful to Waitrose, the one supermarket that undertakes serious missionary work among us *pagani* of rural England. Two years ago, in the Marlborough branch, I came across bundles of long green stalks wrapped – like orchids – in cellophane cones. At their tips were buds, some tight, some showing a hint of white petals.

I brought them home, snipped them into the juices of roast lamb, and into buttered green beans. Delicious, and as with our mauve chive flowers, those buds had quite a strong flavour. In all they were stronger. It's a chive I should dearly like to grow.

In the East you will find the leaves of other cultivars of this species, which have been blanched in picturesque style under pairs of curved roof tiles or in tunnels of straw mats. They may come as a delicate garlicky touch in stir-fried dishes, with *chow mien*, or floating in soups.

GALANGAL

(Alpinia galanga)

On account of medieval recipes I would suspect that the words galantine and galangal are confused in the inquiring cook's mind. Galangal was used to flavour a galantine sauce, which consisted of bread crusts, a little galangal, cinnamon and ginger, pounded up and moistened with an appropriate stock. This was heated, sharpened with a splash of vinegar and strained over fish or meat, a dashing lively sauce, a gallant sauce:

> Was never pike wallowed in galantine
> As I in love am wallowed and y-wound

as Chaucer wrote, round about AD 1384.

The word galangal came via Persian and the Arabic *khalanjan* from the Chinese *Ko-liang-kiang*, meaning no more than mild ginger from Ko, a prefecture in the province of Canton. Europe seems to have stopped using galangal with the development of simpler cooking, or rather simpler flavouring, in the Renaissance. I doubt it appears at all in the seventeenth-century cookery books that set our modern style.

Like ginger, it is the rhizomes of galangal which are used. They are harder and have a slightly different, if gingery flavour, but they are prepared in the same way. Galangal is often part of the spice mixtures in south-east Asian and southern Chinese cookery. Alan Davidson gives a delicious paste for flavouring crab meat, along with a number of other recipes using galangal, in his *Seafood of South-east Asia*: crush and chop enough lemon grass to make 3 level tbsps. Crush to a paste with 1½ level tsps chopped Florentine fennel, 1 shallot and 4 cloves garlic all chopped, 1 level tsp each chopped galangal and chopped Kaffir lime peel (or substitute lime or lemon peel) and 2 tsps shrimp paste. Fry the mixture golden brown in a very little oil. Add 1 cup thick coconut milk (see p. 39) and a splash of *nuoc mam* (fish sauce). Mix this with the meat from 6 boiled crabs, adding extra coconut milk to taste, and seasoning. Fill the crab shells with this mixture and bake 15 minutes. Scatter with chopped fennel leaves and serve.

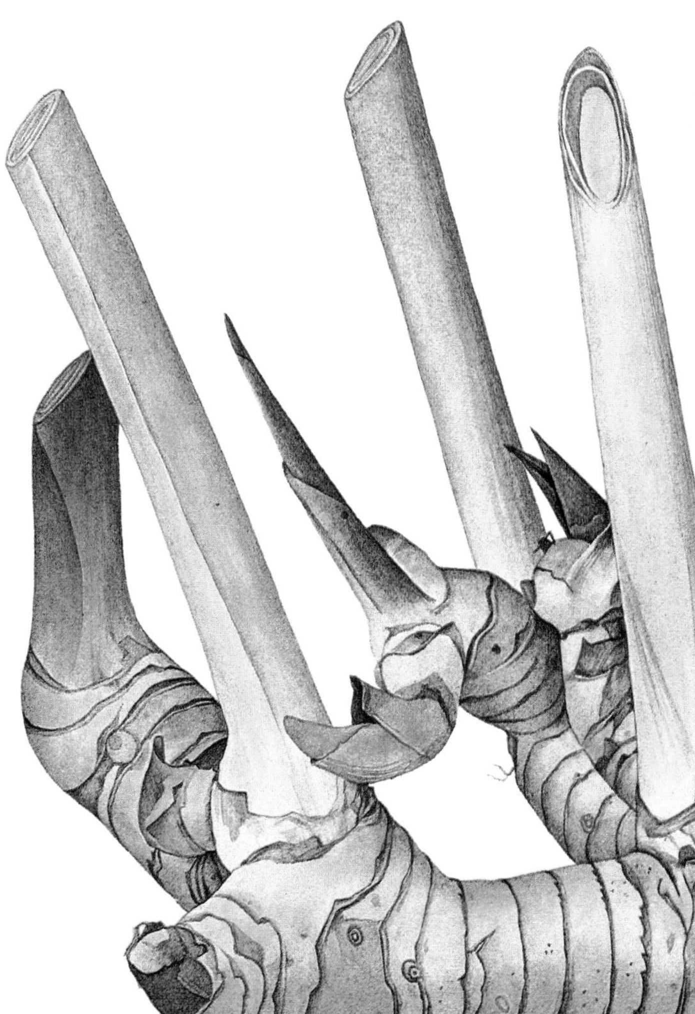

HOT PEPPER

(Capsicum sinensis)

I was once preparing dinner in a hurry and left out two small dishes of cooked pear to cool down on top of the washing-up machine near the sink. The next job was cleaning sweet and hot peppers for a Lebanese dish of grilled chicken. I scraped out the seeds under the tap then cut them in pieces on the draining board, and completed all the preparations. We sat down to the chicken, which was fine, and my husband started on his pears. 'What a good idea', he said, 'to pepper them! It brings out their flavour.' I hadn't peppered them. Trying to think of an explanation, I rubbed my eyes – and soon felt the most painful burning sensation. There was the answer. The fiery capsicin, particularly of the hot peppers, must have wafted from the sink on to the fruit, as well as into my skin. Take warning from this – prepare hot chillies under the tap wearing rubber gloves and rinse any utensils – particularly the goblet of blender or processor – several times with special care.

I remember reading once that there are over 300 varieties of capsicum, from the three species *C. annuum* (all the sweet peppers and some of the fiery ones), *C. frutescens* (tabasco and other small tapering peppers, very pungent) and *C. sinensis* (which includes the Scotch bonnet forms, which are so highly prized in Jamaica, as well as some of the very long thin kinds).

There is no sure-fire way of telling how hot any of the pungent kinds is going to be. As a general rule the thinner they are, the hotter. Remember that the seeds tend to be hotter than the fleshy part (this of course applies particularly to sweet peppers), and that they should be removed unless a whole hot pepper is required. Sri Owen in her *Indonesian Food and Cookery* remarks that she did not dare give the quantities normal to her compatriots' taste: and I imagine that this is also the case with other books on south-east Asian cookery written for Westerners. Should a friend from those parts of the world invite you to a meal, do not be too insistent on your own desire to eat 'real' south-east Asian food: you may well not be able to enjoy a mouthful unless your hostess has made concessions to Western sensitivity, and may end up spluttering and in tears like Becky Sharp when she imprudently took a mouthful of chillies, in *Vanity Fair*, thinking they would cool her down after the peppery curry.

Here is a fricassée of chicken from Jamaica, from Norma Benghiat, showing how you might use hot peppers (they are also known by the more evocative names of Granny's bonnet, bonnet pepper or Scotch bonnet). Rub a 2 kg (4 lb) dressed chicken with lime juice and joint it. Season well. Make a marinade of 1 crushed clove garlic, the leaves of a sprig of thyme, 2 sliced onions, 2 chopped tomatoes and 4 slices hot red pepper minus the seeds. Turn the pieces over so that they are covered with the marinade. Tuck in a whole green hot pepper. Leave 15 minutes at least.

Remove the pieces from the marinade, scraping them clean. Fry until well browned in a little oil, then remove them to a plate. Pour off any fat remaining in the pan, and deglaze it with the marinade mixture, scraping the bottom free. Add ½ l (18 fl oz) water and bring to the boil. Put back the chicken, cover and simmer until the bird is tender. Do not break the whole pepper, or the seeds will come out and make the dish hotter than you had bargained for, and be careful to remove it just before serving. Fried plantain, avocado, a salad and a dish of rice with peas are usually put on the table as well, turning this fricasséed chicken into a magnificent meal for Sundays or some celebration.

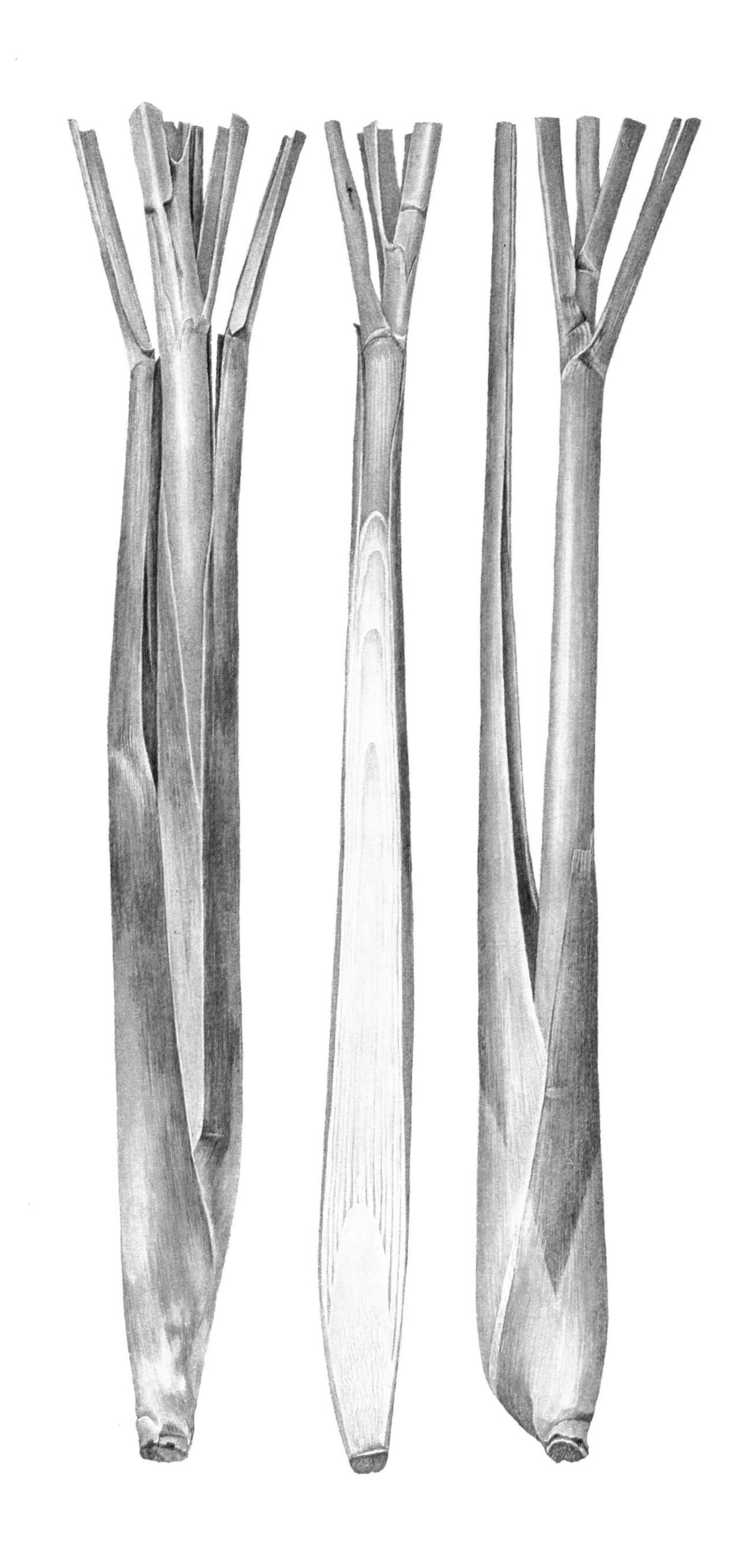

LEMON GRASS

(Cymbopogon citratus)

Lemon grass is easily recognized from its tufts of long grass-like leaves. When you encounter it at the market it has been trimmed at the top, like a head of celery, and the coarsest outer leaves removed. Some recipes will specify the white inner part of lemon grass, others the green part. Or you will just be required to use two stalks or leaves, which should then be crushed slightly and chopped. The outer leaves can be used as flavouring then, like bay leaves, removed before the dish is served. You may find it easier to get dried lemon grass, or lemon grass powder, which should be used by the pinch.

Lemon grass is one of the flavourings that seem to be gaining popularity in the West with young chefs working in the new style. I recall one particularly delicate fish soup at Hubert's restaurant in New York, and being told that the chef – originally from the Philippines – had used lemon grass in the stock. Here is a Laotian fish soup which is not dissimilar.

Keng som pa

Scale and cut into 3 cm (1 in) sections, 500 g (1 lb) of a good freshwater fish. Put into a pan 3 stalks of lemon grass, crushed, a finger-thick slice of ginger and 6 slices of galangal which have been roasted in the oven, with ½ tsp each of salt and monosodium glutamate. Add ¾ l (27 fl oz) water and boil gently for 10 minutes. Add the fish and 2 tbsps *nam pa* (fish sauce). Bring back to the boil and add either 2 small tomatoes, quartered, or slices of peeled green mango, or a slice of peeled pineapple cut in neat wedges. Simmer for a further 10–15 minutes, uncovered. Remove from the heat, discard lemon grass, ginger and galangal pieces. Add 3 finely sliced spring onions and 2 tbsps chopped green coriander. Add a few drops of lime juice to the soup, then distribute equitably.

Sometimes lemon grass will be used with other acidities, tamarind for instance, and even lemon juice or lime as above. On other occasions it will just be one item in a general spice blend as in this curry.

Sumatra Lamb Curry

Cut 1 kg (2 lbs) boned shoulder of lamb into convenient cubes. Mix together 6 long red chillies, seeded but left whole, 1 tbsp shrimp paste *(blachan)*, 6 large chopped shallots or 4 medium onions, 3 chopped cloves garlic, ½ tsp ground turmeric, 2 tsps chopped lemon grass, a peeled chopped slice of galangal and 8 chopped macadamia nuts.

Heat enough oil to make a thin layer in a heavy sauté pan and stir-fry the spice mixture until it begins to smell aromatic, then add the lamb and turn it over well. Pour in ½ l (18 fl oz) coconut milk (see p.39), then enough water just to cover the meat. Simmer until the meat is tender. If you can find pea aubergines (p.116), add 8 or 10 to the dish and give it another 5 minutes. Alternatively add a small aubergine, cut into roughly the same size chunks as the meat. Dish up the meat into a bowl, removing any fat floating on top of the sauce. Taste it for seasoning (you can give it a slight extra sharpening of tamarind water if you like) and pour over the meat. Arrange 1 or 2 of the red chillies on top. Serve with rice, and perhaps a cucumber sambal.

PEA AUBERGINE

(Solanum torvum)

If you walk along the sunny edge of some tropical forest in Indonesia or Thailand, or by the shadowed bank of a river or canal, you will see pea aubergine growing wild. Nothing more than a common weed, although the small bitter fruits are so popular a flavouring that it is also cultivated. In the market clusters of these mirabelle-sized aubergines, going from green to what one botanist described as 'sordidly yellow', are sold in large, tightly packed bunches – an indication of popularity since you rarely need more than six to eight for any dish. In south-east Asia they go raw into sauces and salads, but in Jamaica – where they are known as susumbers or gully beans – they are cooked and added to dishes of salt cod.

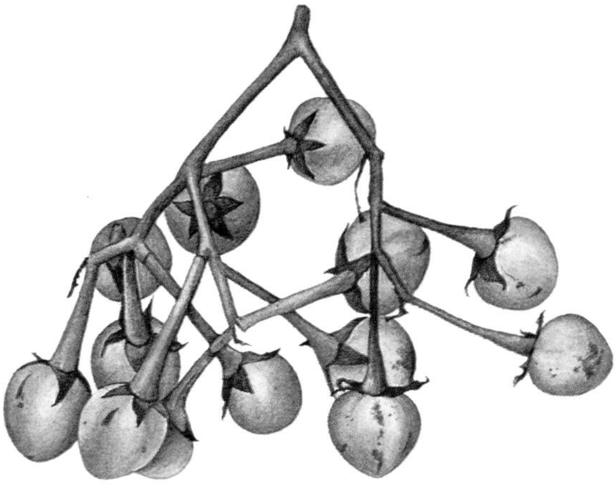

SCREWPINE

(Pandanus amaryllifolius)

In the West we forget the close relationship between perfume and flavourings. Odd to think that Parisian perfumiers once put it about that vanilla was poisonous to halt its increasing use by pastrycooks. Pandanus or screwpine species provide scented flowers for kewra essence to flavour syrupy Indian desserts, and scented leaves for cookery, for medicine and for religious offerings.

The leaves of the small screwpine illustrated give a delicate green colour and an air of young rice or new-mown hay to custards and rice dishes. In Indonesia, on the island of Alor, Sri Owen came across a steamed cake – *serikaya* – flavoured with 2 small bits of screwpine leaf put into the base of the tin (16 cm or 6½ in). For the batter, beat 3 large eggs, add 90 g (3 oz) brown sugar, beat again well and add 125 g (4 oz) ground almonds. Steam for 30 minutes, and eat hot or cold after removing the leaves.

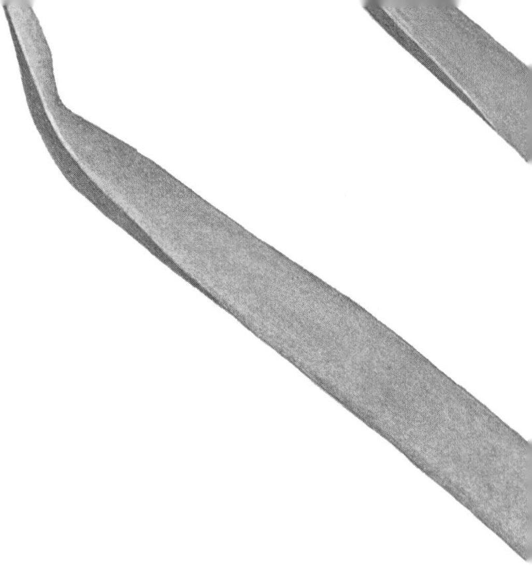

SEBESTEN

(Cordia myxa)

In his *Inquiry into Plants,* Theophrastus the botanist and pupil of Plato and Aristotle, wrote

> there is another tree, the Egyptian plum, which is of great size, and the character of its fruit is like the medlar which it resembles in size, except that it has a round stone. It begins to flower in the month of October, and ripens its fruit about the winter solstice and it is evergreen. The inhabitants of the Thebaid, because of the abundance of the tree, dry the fruit; they take out the stones, bruise it, and make cakes of it.

And I suppose that it was in this handy condition that it entered into European medicine, its mucilaginous and slightly astringent pulp being supposedly soothing to bad chests. The word sebesten appears in various European languages – English, French, Spanish and Portuguese – in late medieval medical texts. It was still being recommended for a soothing syrup for asthma in the last century, together with dates, raisins and jujubes. The intriguing name is thought to come from the Arabic, *seg pistan,* meaning dog's teats, though it is difficult to see why this should be so.

The plummy, prune-like nature of dried sebesten is not at all evident in the green unripe fruit you see today in Asian markets. When Charlotte Knox first saw the clusters of sebesten she could not discover their identity. In the end she went along to the herbarium of the Natural History Museum, and the botanist there was stuck as well. At least until she had the bright idea of showing it to an Indian warden. He was excited to see it and told her what it was. He calls it gunda rather than sebesten, and eats it pickled – with *dal* and other slightly boring foods that need a bit of a lift. It can also be eaten when ripe, in a raw or cooked state. It is at this stage that it will be closer to Theophrastus's medlar, which when fully ripe and brown – bletted is the correct word – has date-like qualities, something we do not find very attractive today but which must have been more appreciated in the past in winter when fruits were scarce and sweetness was longed for.

SUGAR CANE

(Saccharum officinarum)

Sugar cane is a giant grass that was first cultivated in India in prehistoric times. By about 500 BC they had even devised a machine for crushing out the juice, which suggests that sugar was being made. A couple of hundred years later sugar had reached the Mediterranean. Pliny knew all about it when he was writing his *Natural History* in the first century AD: 'It is a sort of honey that collects in reeds, white like gum and brittle to the teeth, the biggest bits are the size of hazelnuts. It is only used as a medicine.'

And as medicine it continued, often in the expensive form of rose- and violet-flavoured sweets and comfits (seeds dipped in many coatings of sugar). Black teeth were the sign of privilege in Tudor England. Not until we developed a slave trade to exploit the sugar plantations of our West Indian colonies did black and gappy teeth become available to all. These days our average consumption is reckoned to be something like 5·75 heaped tablespoonsful of sugar a day.

Of course we – and the Americans – are not the only people with a sweet tooth. Tribes living in the highlands of New Guinea, tribes undiscovered until the 1930s, developed many different cultivated varieties of sugar cane in spite of the limitations of their Stone Age culture. They selected the juiciest canes, the ones with the least central fibre and the prettiest appearance, to satisfy what seems to be an instinctive human passion for sweetness.

When you buy a bit of cane, you pull off the stripy rind with your teeth, then chew at the fibrous centre, sucking out the juice. Then you spit out the exhausted residue.

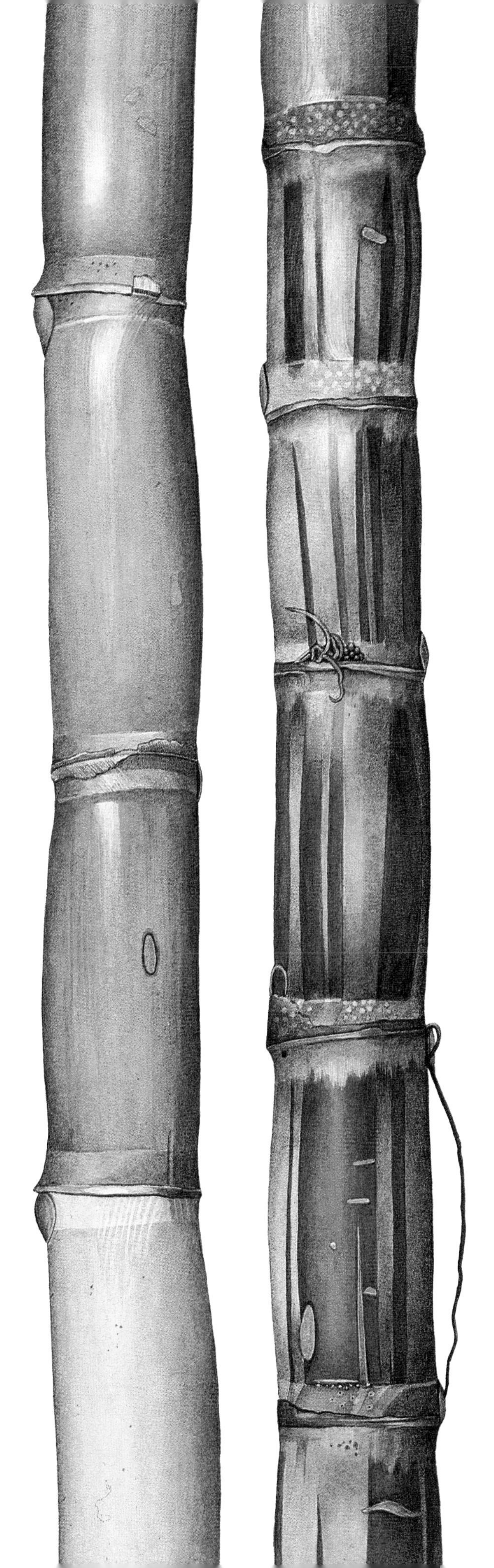

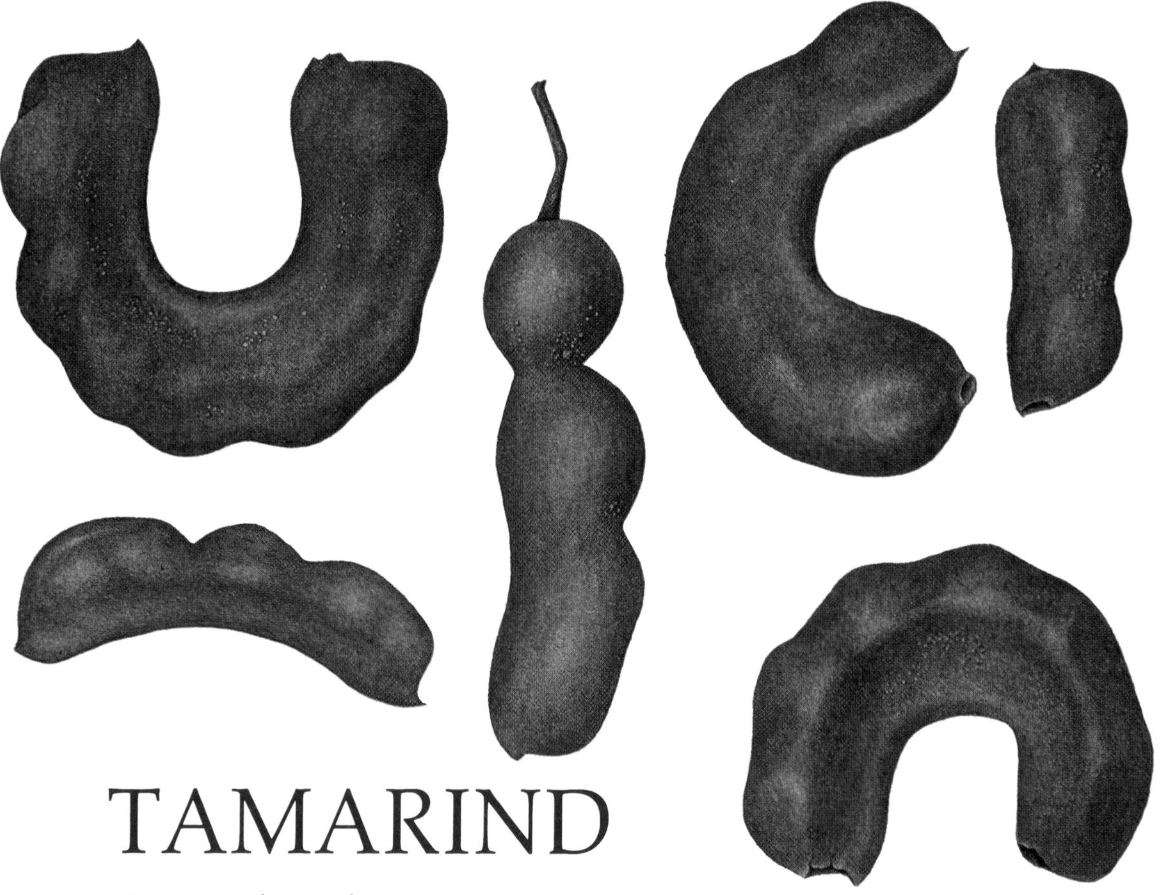

TAMARIND

(Tamarindus indica)

The name tamarind, from the Arabic *tamr-hindi* meaning Indian date, gives you an idea of the colour and consistency of the edible pulp that surrounds the seeds and fills the pods of this fine and beautiful tree. The flavour, though, is quite different from a date's. It has a sweet acidity that is gentler and more refreshing than anything that citrus fruit can provide. Nothing quite takes its place in oriental cookery.

Usually we have to buy tamarind from Asian stores in plastic-wrapped blocks. It is dark-brown and messy, uneven with seeds, and you just have to break off what you want and cope as best you can with the soaking and loosening of the pulp. In the markets you may have the luck to find tamarind pods like the ones in the illustration. Crack the brittle, dark-brown, scurfy-looking pod and scrape out the doughy brown pulp and seeds. The pulp can be eaten as it is, or mixed with sugar and rolled into sweetmeats.

More deliciously, make this chutney-like sauce *soonth* (which means ginger): soften 125 g (4 oz) tamarind, minus the pod, in 600 ml (1 pt) water for about half an hour. With your fingers break up the mass and pull out any seeds that can be cleanly removed. Meanwhile toast 1 rounded tsp cumin seeds in a little pan over a moderate heat until they begin to darken. Add to the pan with 150 g (5 oz) soft brown sugar, 1 rounded tsp ground dry ginger and a hot seeded chilli. Stir over a low heat until the sugar is dissolved, then boil to a chutney-like consistency. Sieve, add a handful of sultanas, bring to the boil again and store in a jar. Serve with crisp *gol gappas* or *papri*, or Western-style with cold chicken as a chutney.

To make a quick fresh chutney, soak 30 g (1 oz) seedless tamarind pulp with 5 tbsps water for 20 minutes. Grind 30 g (1 oz) each coriander and mint leaves with a seeded green hot chilli. Mix in the tamarind water, 1½ level tsps sugar and 1 level tsp salt. Serve with cooked pulses or *dals*.

TURMERIC

(Curcuma longa)

Turmeric is a plant of India and south-east Asia, the part used as a spice being the bright orange–yellow rhizome. Although you can buy it fresh in its native lands, the kind we get has been boiled, peeled and dried in the sun for several weeks until it is very hard indeed. For this reason it is almost always sold ready ground. I have seen some writers suggesting turmeric powder as an alternative to saffron, for instance in Spanish dishes. As far as food is concerned it is no such thing (though I concede that as a dye it produces convincing 'saffron' robes). It has its own special place in cookery. The bright colour in piccalilli and mustards and commercial curry powders comes from turmeric. More happily it appears in Indian dishes of vegetables or rice, like the two following.

Sesame Rice

Cook 500 g (1 lb) basmati or patna rice in salted water and drain. Mix in 3 tbsps sesame oil. Mix together 1 tsp each of turmeric, fenugreek, chilli or cayenne pepper, with 2 tsps each coriander seeds, sesame seeds and *urhad* (black) *dal*, and 1 tbsp gram *dal*. Cook them in 2 tbsps sesame oil for a minute or two, stirring all the time, then stir into the rice. In a larger pan cook a rounded tbsp of skinned peanuts, 2 tsps mustard seeds and 6 curry leaves in a further 2 tbsps of sesame oil for a few seconds until the seeds begin to pop, then quickly stir in the rice. Add lemon juice to taste.

Avial

A splendid mixed dish of vegetables. Assemble a good kg (2 lbs) of vegetables, cutting them into 1 cm (½ in) cubes or slices as appropriate. Use potato or sweet potato, a large green banana, an onion and an aubergine if possible, then add whatever else you may have – carrots, green or broad beans, sprouts, celery, fennel, pumpkin, courgette or other squash, or white radish. Cook the heftier vegetables first in boiling salted water, with 1 tsp cumin, 1½ tsps turmeric and 6–8 curry leaves. Add the remaining vegetables according to their tenderness.

When the vegetables are almost cooked, stir in 2 tbsps tamarind water (p. 120), the grated meat of half a coconut and simmer for 5 minutes. Meanwhile beat up 250 ml (8 fl oz) good quality yoghurt. Take the pan from the heat, stir in the yoghurt and taste for seasoning. Turn it into a hot dish rather than putting it back on to the stove (the yoghurt could curdle).

Serve with the sesame rice and poppadums. These are two festival dishes of southern Indian cookery – in those parts a good number of red hot chillies would be sliced and stirred into the sesame rice.

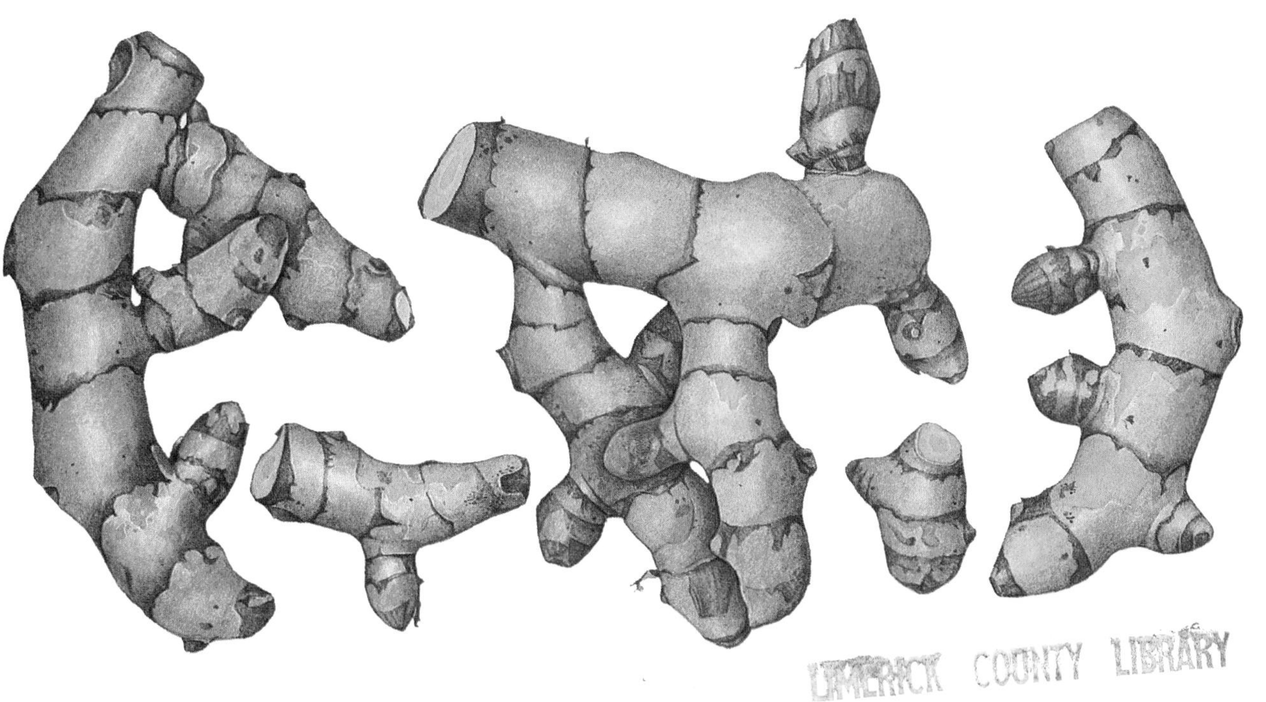

GLOSSARY

Fruits, vegetables, herbs and spices have been listed in the text under the names they are known by to botanists. However, they also have a number of common or regional names by which they are known and which your local shop or market might use. These are listed below.

Angled loofah (*Luffa acutangula*)	ridged or club gourd; sze gwa, 'silk gourd' (China); jhinga-torooee, jhingli, kali, turia (India); oyong (Indonesia); petola (Malaysia, Sulawesia)
Asparagus bean	*see* winged bean
Banana flower or heart (*Musa* sp.)	jantung pisang (Indonesia); djantoong (Malaysia); hua blee (Thailand)
Banana leaf (*Musa* sp.)	
Bean	*see* asparagus bean, cluster bean, cowpea, hyacinth bean, pigeon pea, sa-taw bean, yard-long bean
Betel (*Piper betle*)	pan (India)
Bitter gourd (*Momordica charantia*)	balsam pear, bitter cucumber, leprosy pear; foo gwa (China); karela, karella, kerela (India); peria (Indonesia); tsurureishi (Japan)
Black-eyed bean	*see* cowpea
Bottle gourd (*Lagenaria siceraria*)	calabash gourd, do-di, dudi, white-flowered gourd; woo lo gwa (China); lokhi (India)
Bread-fruit (*Artocarpus altilis*)	
Carambola (*Averrhoa carambola*)	star fruit; kamrakh, karmara (India); belimbing manis (Indonesia); mafueng (Thailand)
Chayote (*Sechium edule*)	custard marrow, pepinello, vegetable pear; faat sua gwa, 'Buddha's hand gourd' (China); mirliton (Louisiana); chaco, chocho, choko, chow-chow, christophine (West Indies)
Cherimoya (*Annona cherimolia*)	
Chinese key (*Boesenbergia pandurata*)	kachai (Thailand)
Cluster bean (*Cyamopsis tetragonoloba*)	guar (India)
Cowpea (*Vigna unguiculata* subsp. unguiculata)	black-eyed bean, black-eyed pea, chola, chopia
Curry leaf (*Murraya koenigii*)	gandla, katnim, mitha neem (India); daun salaam (Indonesia); bai karee (Thailand)
Cush-cush (*Dioscorea trifida*)	aja, aje, yampee
Dasheen leaf (*Colocasia esculenta* var. esculenta)	elephant's ear; bhaji, patra (India); daun talas (Indonesia); callaloo (West Indies)
Drumstick (*Moringa oleifera*)	morunga, sahajna (India)
Fingered yam	*see* greater Asiatic yam
Flowering Chinese chive (*Allium tuberosum*)	Cuchay or Cantonese onion, fragrant onion, spring flowers; gau choi fa (China)
Galangal (*Alpinia galanga* syn. *Languas galanga*)	galingale, Siamese ginger; laos, lengkuas (Indonesia); ka (Thailand)

Gourd	*see* angled loofah, bitter gourd, bottle gourd, ivy gourd, smooth loofah, snake gourd, spiny bitter gourd
Greater Asiatic yam (*Dioscorea alata*)	greater yam, water yam, white Manila yam; cultivars include Barbados yam, fingered yam, soft yam and white Lisbon yam
Guava (*Psidium guajava*)	
Hot pepper (*Capsicum sinensis*)	bonnet pepper, granny's bonnet, Scotch bonnet
Hyacinth bean (*Lablab purpureus*)	bovanist bean, Egyptian kidney bean, lablab bean, papaya bean
a) Papri (*Lablab purpureus* subsp. purpureus)	popat, popetti (India)
b) Seim (*Lablab purpureus*)	
c) Valour (*Lablab purpureus* subsp. bengalensis)	val, valla, valor, wal (India)
Ivy gourd (*Coccinia grandis*)	gelowda, kundree, ole kavi, telacucha, tindola, tindori (India)
Kiwi (*Actinidia chinensis*)	Chinese gooseberry
Lemon grass (*Cymbopogon citratus*)	citronella grass, geranium grass; heung maau ts'o, 'fragrant thatch grass' (China); herva chaha, khawi (India); serah, sereh (Indonesia); takrai (Thailand)
Lesser Asiatic yam (*Dioscorea esculenta*)	Chinese yam, fancy yam, potato yam, sweet yam; chini alu, 'sugar yam' (India)
Lychee (*Litchi chinensis*)	litchee, litchi
Mango (*Mangifera indica*)	
Mangosteen (*Garcinia mangostana*)	
Okra (*Hibiscus esculentus*)	bamia, bamies, gumbo, lady's fingers, okro; bindi (India)
Passion fruit (*Passiflora edulis*)	purple granadilla
Pawpaw (*Carica papaya*)	papaw, papaya
Pea aubergine (*Solanum torvum*)	gully bean, susumber (Jamaica); suzume-nasubi, 'sparrow aubergine' (Japan); makeua puong (Thailand)
Persimmon (*Diospyros kaki*)	kaki
Pigeon pea (*Cajanus cajan*)	Congo bean, no-eye pea, red gram; arhar, kandi, rahar, tur, tuver (India); goon-goon, gungar, gungo (West Indies)
Plantain (*Musa paradisaica*)	kayla (India)
Pomegranate (*Punica granatum*)	
Prickly pear (*Opuntia ficus-indica*)	Barbary fig, cactus berry, cactus fig, Indian fig, tuna
Rambutan (*Nephelium lappaceum*)	
Roselle (*Hibiscus sabdariffa*)	Jamaican sorrel, red sorrel, rosella, sorrel, sour-sour
Sapodilla (*Manilkara zapota* syn. *Achras sapota*)	chiku, dilly, naseberry, sapota, sapodillo, zapota
Sa-taw bean (*Parkia speciosa*)	petai, peté, peteh (Indonesia); sa-taw (Thailand)

Sugar apple or custard apple
(*Annona squamosa*)
(see p. 14)

Screwpine (*Pandanus amaryllifolius* syn. *Pandanus odoratus*)	kewra, pandanus; daun pandan (Indonesia); bai toey hom (Thailand)
Sebesten (*Cordia myxa*)	gunda, scilla (India)
Smooth loofah (*Luffa cylindrica*)	African sponge, dish cloth gourd, sponge loofah, sponge towel gourd, vegetable sponge; seui gwa (China); dhundal, jhinga, mozhuku peekankai, turai (India)
Snake gourd (*Tricosanthes cucumeria* var. anguina)	serpent cucumber, serpent gourd, viper gourd; chichinda, podalangai (India)
Sour-sop (*Annona muricata*)	guanábana, prickly custard apple
Spiny bitter gourd (*Momordica cochinchinensis*)	spiny bitter cucumber; kakola, kakrol, kakur, kantola (India); fak kao (Thailand)
Sugar apple (*Annona squamosa*)	custard apple, sweet-sop; sharifa (India)
Sugar cane (*Saccharum officinarum*)	noble cane
Sweet potato (*Ipomoea batatas*)	faan sue (China); Louisiana yam (southern USA); yam (USA); iyam, nyam (West Africa)
Sweet yam	*see* lesser Asiatic yam
Tamarind (*Tamarindus indica*)	Indian date; asam (Indonesia); mak kam (Thailand)
Taro (*Colocasia esculenta*)	woo tau (China); colocassi, colokas, koloka, kolokasia, kolokassi (Cyprus); arvi (India); talas (Indonesia); sato-imo (Japan); gra dat, puak (Thailand); baddo, coco, cocoyam, dasheen, eddoe (West Indies)
Turmeric (*Curcuma longa*)	wong keung, 'yellow ginger' (China); halad, haldie, huldie (India); kunyit (Indonesia); ukon (Japan); kamin (Thailand)
White radish (*Raphanus sativus* var. longipinnatus)	Chinese radish, moulli, oriental radish; loh baak (China); daikon (Japan)
White yam (*Dioscorea rotundata*)	Brazil yam, eboe yam, eight months' yam, white Guinea yam
Winged bean (*Psophocarpus tetragonolobus*)	asparagus bean, asparagus pea, four-angled bean, Goa bean, Manila bean, princess pea; sz kok tau (China); kecipir (Indonesia); shikaku-mame, 'square bean' (Japan); tua poo (Thailand). *See also* yard-long bean
Yam	*see* cush-cush, fingered yam, greater Asiatic yam, lesser Asiatic yam, white yam, yellow yam
Yard-long bean (*Vigna unguiculata* subsp. sesquipedalis)	asparagus bean, asparagus pea, bodi bean, Chinese bean, snake bean; dau gok (China); kacang panjang (Indonesia); jūroku-sasage (Japan); tua fak yaw (Thailand); boonchi (West Indies)
Yellow yam (*Dioscorea cayenensis*)	affou yam, twelve months' yam, yellow Guinea yam

BIBLIOGRAPHY

Aside from the standard books that are easily available and well known – by Claudia Roden, Elisabeth Lambert Ortiz (to whom I owe a special debt), Madhur Jaffrey and Kenneth Lo, Peter and Joan Martin – I have consulted the following for the main part of my information:

Anthonio, H. O., and Isonu, M., *Nigerian Cookbook* (Macmillan, London, 1982).
Arends-Savelkeul, L., van der Sar-Vuyk, J. M., and Senior-Baiz, E., *Cookbook of the Netherlands Antilles* (Boekhandel Salas, Curaçao, 1968).
Balbir Singh, Mrs, *Indian Cookery* (Mills & Boon, London, 1961).
Bazore, Katherine, *Hawaiian and Pacific Foods* (M. Barrows, New York, 1940).
Benghiat, Norma, *Traditional Jamaican Cookery* (Penguin, London, 1985).
Bhatia, Savitri, *Shahi Tukre* (Wheeler, Allahabad, 1975).
Biarnès, Monique, *La Cuisine Sénégalaise* (Société Africaine d'Edition, Dakar, 1972).
Bley Miller, Gloria, *Thousand Recipe Chinese Cookbook* (Hamlyn, London, 1966).
Blue, Betty A., *Authentic Mexican Cooking* (Prentice-Hall, New Jersey, 1977).
Brennan, Jennifer, *Thai Cooking* (Jill Norman, Hobhouse, London, 1981).
Brothwell, Don and Patricia, *Food in Antiquity* (Thames & Hudson, London, 1969).
Burkill, I. H., *A Dictionary of Economic Products of the Malay Peninsula* (Crown Agents, London, 1935).
de Candolle, Alphonse, *Origin of Cultivated Plants* (Kegan Paul, Trench, London, 1886).
Couffignal, Huguette, *The People's Cook Book* (Macmillan, London, 1978).
Coursey, D. G., *Yams* (Longman, London, 1967).
Davidson, Alan, *Mediterranean Seafood* (Penguin, London, 1972).
——, *Fish and Fish Dishes of Laos* (Imprimerie Nationale Vientiane, 1975).
——, *Seafood of South-east Asia* (Author, London, 1976).
Fuller, David, *Maori Food and Cookery* (Reed, Wellington, 1978).
Grey, Winifred, *Caribbean Cookery* (Collins, London, 1965).
Herklots, G. A. C., *Vegetables in South-east Asia* (Allen & Unwin, London, 1972).
Husain, L., *Muslim Cooking of Pakistan* (Ashraf, Lahore, 1974).
The Indian Cookery Book, by a Thirty-five Years' Resident (Wyman, Calcutta, n.d.).
Johns, L., and Stevenson, V., *The Complete Book of Fruit* (Angus & Robertson, London, 1979).
Khaing, Mi Mi, *Cook and Entertain the Burmese Way* (Karoma, Ann Arbor, 1978).
Liberty Hyde Bailey Hortorum, *Hortus Third* (Macmillan, New York, 1976).
Majupuria, Indra, *Joys of Nepalese Cooking* (Devi, Gwalior, India, 1980–1).
Mehta, Jeroo, *101 Parsi Recipes*, 2nd edn (Dolphin Press, Colaba, 1979).
Miller, Carey D., Bazore, Katherine, and Robbins, Ruth C., *Some Fruits of Hawaii* (Hawaii Agr. Experiment Station, Honolulu, 1936).
Olaore, Ola, *The Best Kept Secrets of West and East African Cooking* (Foulsham, London, 1980).
Organ, John, *Gourds* (Faber, London, 1963).
Owen, Sri, *Indonesian Food and Cookery* (Prospect, London, 1980).
Pendaeli-Sarakikya, E., and Blaser, Sister A., *Recipe Book for Tanzania* (Macmillan, London, 1965).
Popenoe, Wilson, *Manual of Tropical and Sub-tropical Fruits* (Hafner Press, New York, 1948 edn).
Sahni, Julia, *Classic Indian Cooking* (Morrow, New York, 1980).
Santa Maria, Jack, *Indian Vegetarian Cookery* (Rider & Co., London, 1973).
Scott, David, *The Japanese Cookbook* (Granada, London, 1981).
Sing, Phia, *Traditional Recipes of Laos* (Prospect, London, 1981).
Slater, Mary, *Cooking the Caribbean Way* (Hamlyn, London, 1965).
Sobhana, Princess Rasmi, *The Cambodian Cookbook* (USIS, Phnom Penh, n.d.).
Sokolov, Raymond, *Fading Feast* (Farrar Straus & Giroux, New York, 1979).
Stobart, Tom, *The Cook's Encyclopaedia* (Batsford, London, 1980).
Taneja, Meera, *The Indian Epicure* (Mills & Boon, London, 1979).
——, *New Indian Cookery* (Fontana, London, 1983).
Taw Kritakara, M. L., and Pimsai Amranand, M. R., *Modern Thai Cooking* (Duang Kamol, Bangkok, 1977).
Tay, Gilda, *Indonesian and Malaysian Cooking* (Bay Books, Sydney, 1978).
Toupin, Elizabeth Ahn, *Hawaii Cookbook and Backyard Luau* (Silvermine, Norwalk, Conn., 1967 edn).
Wickramasinghe, Priya, *Oriental Cookbook* (Sphere, London, 1983).
Yen Hung Feng, Doris, *The Joy of Chinese Cooking* (Faber, London, 1952).
Yule, Col. Henry, and Burnell, A. C., *Hobson-Jobson* (Munshiram Manoharlal, Delhi, 1903, 1968 edn).

Index